M3

B4

Graphs on surfaces

This publication forms part of an Open University course. Details of this and other Open University courses can be obtained from the Student Registration and Enquiry Service, The Open University, PO Box 197, Milton Keynes, MK7 6BJ, United Kingdom: tel. +44 (0)870 333 4340, e-mail general-enquiries@open.ac.uk

Alternatively, you may visit the Open University website at http://www.open.ac.uk where you can learn more about the wide range of courses and packs offered at all levels by The Open University.

To purchase a selection of Open University course materials, visit the webshop at www.ouw.co.uk, or contact Open University Worldwide, Michael Young Building, Walton Hall, Milton Keynes, MK7 6AA, United Kingdom, for a brochure: tel. +44 (0)1908 858785, fax +44 (0)1908 858787, e-mail ouwenq@open.ac.uk

The Open University, Walton Hall, Milton Keynes, MK7 6AA.

First published 2006.

Copyright © 2006 The Open University

All rights reserved; no part of this publication may be reproduced, stored in a retrieval system, transmitted or utilised in any form or by any means, electronic, mechanical, photocopying, recording or otherwise, without written permission from the publisher or a licence from the Copyright Licensing Agency Ltd. Details of such licences (for reprographic reproduction) may be obtained from the Copyright Licensing Agency Ltd, 90 Tottenham Court Road, London W1T 4LP.

Open University course materials may also be made available in electronic formats for use by students of the University. All rights, including copyright and related rights and database rights, in electronic course materials and their contents are owned by or licensed to The Open University, or otherwise used by The Open University as permitted by applicable law.

In using electronic course materials and their contents you agree that your use will be solely for the purposes of following an Open University course of study or otherwise as licensed by The Open University or its assigns.

Except as permitted above you undertake not to copy, store in any medium (including electronic storage or use in a website), distribute, transmit or re-transmit, broadcast, modify or show in public such electronic materials in whole or in part without the prior written consent of The Open University or in accordance with the Copyright, Designs and Patents Act 1988.

Edited, designed and typeset by The Open University, using the Open University T$_E$X System.

Printed and bound in the United Kingdom by The Charlesworth Group, Wakefield.

ISBN 0 7492 4132 2

1.1

Contents

Introduction		**4**
	Study guide	5
1	**Colouring maps on a sphere**	**6**
	1.1 Colouring maps on a plane	7
	1.2 Colouring maps on surfaces	8
	1.3 Euler's Formula	12
	1.4 The Five-colour Theorem	16
2	**Colouring graphs**	**20**
	2.1 Graphs	21
	2.2 Planar graphs	24
	2.3 Colouring planar graphs	27
3	**Embedding graphs on surfaces**	**30**
	3.1 Revisiting surfaces	30
	3.2 The orientable genus	33
	3.3 The non-orientable genus	38
4	**Colouring maps on surfaces**	**41**
	4.1 Maps on a torus and a projective plane	41
	4.2 Maps on orientable surfaces	44
	4.3 Maps on non-orientable surfaces	49
5	**Proving the Four-colour Theorem**	**53**
Solutions to problems		**60**
Index		**64**

Introduction

In *Unit B1* we introduced the idea of a *compact surface* as a polygon with edge identifications and presented several examples of such surfaces, such as the sphere, torus, projective plane and Klein bottle. We explained what is meant by a *subdivision* of a compact surface, and introduced the *Euler characteristic* χ of the surface; this is one of the three numbers that we use to classify compact surfaces, the other two being the *boundary number* β and the *orientability number* ω. You also met briefly the concept of a *connected graph* drawn on the surface.

In *Unit B2* we used the Euler characteristic to determine the regular subdivisions of a given compact surface, and introduced the concept of a *dual subdivision*. Then, in *Unit B3*, we proved that a compact surface is uniquely determined by the three numbers χ, β and ω. We observed that a compact orientable surface without boundary is topologically equivalent to a sphere with a number of handles — for example, a torus can be regarded as a sphere with one handle — and that a compact non-orientable surface without boundary is topologically equivalent to a sphere with a number of cross-caps — for example, a projective plane can be regarded as a sphere with one cross-cap and a Klein bottle can be regarded as a sphere with two cross-caps.

Another way of distinguishing between surfaces is by their map-colouring properties. We saw an example of this in the course DVD, where we stated that any map on a sphere can be coloured with just four colours so that neighbouring countries are coloured differently, whereas maps on a torus may require up to seven colours. In this unit we define a map on a surface to be a certain type of subdivision of the surface, and we investigate the number of colours that are required when we colour maps on the surface. We outline a proof of the *Four-colour Theorem*, which states that four colours are sufficient to colour any map on a sphere, and discuss the *Ringel–Youngs Theorem*, that tells us how many colours are required when we colour maps on other surfaces.

As we shall see, problems involving the colourings of maps on surfaces can frequently be clarified by *dualizing* the problem and investigating the equivalent problem of colouring the vertices of the dual subdivision, which is a connected graph drawn on the surface. Since vertex-colouring does not depend on the particular embedding of the graph on the surface, this can simplify matters considerably. For this reason, we shall interrupt our discussion of map colouring to enter the world of graph theory in Sections 2 and 3, before returning to the colouring of maps in Section 4.

Study guide

As noted above, several of the ideas presented in this unit have already appeared earlier in *Units B1*, *B2* and *B3*, and you may wish to refer back to these units from time to time.

In Section 1, *Colouring maps on a sphere*, we introduce *maps* on a surface as subdivisions of the surface. In particular, we discuss colourings of maps drawn on a plane or a sphere, and we prove that five colours are sufficient to colour every map of this kind.

Colouring the countries of a map drawn on a plane or a sphere is equivalent to colouring the vertices of a planar graph, and in Section 2, *Colouring graphs*, we introduce the language of graph theory and discuss planar graphs as *duals* of the maps in Section 1. We also prove some results on the colouring of planar graphs.

From a topological point of view, the most important sections of this unit are Sections 3 and 4. There are software activities associated with this material.

In Section 3, *Embedding graphs on surfaces*, we extend our discussion of Section 2 to the embedding of graphs on surfaces of arbitrary Euler characteristic. We present various versions of the Ringel–Youngs Theorem for graphs embedded on such surfaces.

In Section 4, *Colouring maps on surfaces*, we return to the colouring of maps on surfaces and present several results for such maps.

Finally, in Section 5, *Proving the Four-colour Theorem*, we outline a proof that four colours are sufficient to colour every map drawn on a plane or a sphere. This section is not assessed.

1 Colouring maps on a sphere

> After working through this section, you should be able to:
> ▶ explain what is meant by a *map* on a surface;
> ▶ explain what is meant by the *colouring of maps on a plane or a sphere*, and state the *Four-colour Theorem*;
> ▶ deduce from Euler's Formula that every map drawn on a plane or sphere has a country with at most five neighbours;
> ▶ outline a proof of the *Five-colour Theorem*.

On 23 October 1852, Augustus De Morgan, Professor of Mathematics at University College, London, wrote a letter to Sir William Rowan Hamilton, the Astronomer Royal of Ireland:

> A student of mine asked me today to give him a reason for a fact which I did not know was a fact — and do not yet. He says that if a figure be anyhow divided and the compartments differently coloured so that figures with any portion of common boundary line are differently coloured — four colours may be wanted, but not more ... Query cannot a necessity for five or more be invented? ... My pupil says he guessed it in colouring a map of England ... The more I think of it the more evident it seems ...

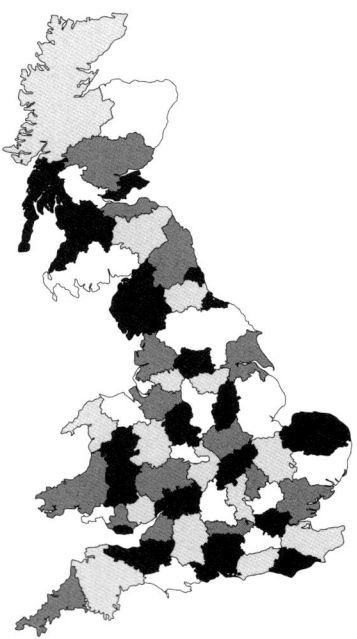

Figure 1.1 Map of mainland Britain coloured with four colours

The problem of whether four colours are sufficient to colour all maps on a plane is known as the *Four-colour Problem*, and for many years it was one of the most famous unsolved problems in mathematics.

In 1879 Alfred Kempe, a London barrister and amateur mathematician, gave a purported solution to the Four-colour Problem, but in 1890 the Durham mathematician Percy Heawood found a serious error in it. Indeed, it was not until 1976 that a proof was found, by two mathematicians at the University of Illinois, Kenneth Appel and Wolfgang Haken, involving the substantial use of a computer.

The problem of colouring maps on a plane is equivalent to that of colouring maps on the surface of a sphere (such as a globe), and Heawood was able to salvage enough from Kempe's attempted solution to prove that *every map drawn on a plane or a sphere can be coloured with at most five colours*. Heawood then showed how to generalize the problem to the colouring of maps on other surfaces, such as a torus, and he presented a formula for the number of colours needed for any surface. Unfortunately, Heawood's argument for justifying his formula was deficient, and his assertion came to be known as the *Heawood Conjecture*. A valid proof was finally obtained in 1968, by two mathematicians at the University of California, Gerhard Ringel and Ted Youngs, and the result is now known as the *Ringel–Youngs Theorem*.

In this section we explain what we mean by a *map*, and we show that colouring maps on a plane is equivalent to colouring maps on the surface of a sphere. We present Euler's Formula for such maps, and use it to prove that every map on a plane or a sphere has at least one country bounded by five or fewer other countries. We deduce that every map on a plane or a sphere can be coloured with at most five colours.

1.1 Colouring maps on a plane

Consider the following colouring of the 48 contiguous states of the USA.

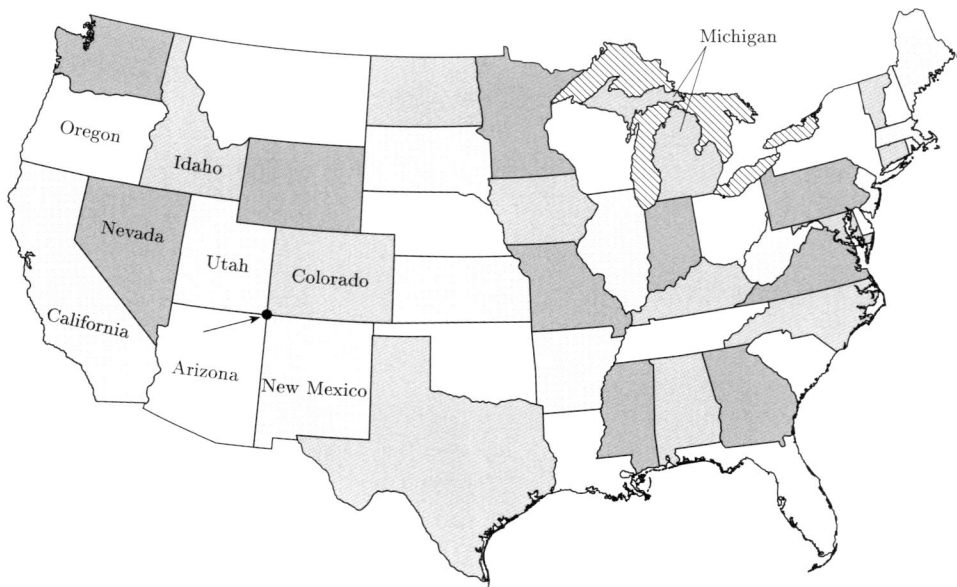

Although not discussed, the five Great Lakes are shown hatched for clarity. Note that the state of Michigan is in two large chunks separated by a narrow channel of water.

Figure 1.2

Four colours are sufficient to colour this map in such a way that any two states with a common boundary line are coloured differently; we say that this map has a *four-colouring*, or is *four-colourable*.

It is natural to ask whether a similar statement can be made for all maps.

By a 'four-colouring', we mean a colouring using *at most* four colours.

> ### Four-colour Problem
> Can the countries of every map drawn on a plane be coloured with at most four colours in such a way that any two countries with a common boundary line are assigned different colours?

Remarks

(i) We often use the word 'countries' for the regions of a map, though they may (as in Figures 1.1 and 1.2) be regions of another type, such as counties or states.

(ii) If two countries meet at a point, rather than along a common boundary line, they can (though they need not) be assigned the same colour. For example, in the United States map (Figure 1.2), four states meet at the point indicated by the arrow; these are (clockwise from top left) Utah, Colorado, New Mexico and Arizona. In the figure, Utah and New Mexico have the same colour. This convention is necessary since, otherwise, the 8-slice pie map in Figure 1.3 would require eight colours; with our convention, only two colours are needed.

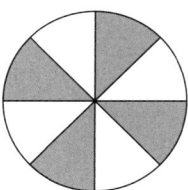

Figure 1.3

(iii) Four colours are needed when a map contains four mutually neighbouring countries. However four colours may be needed even when a map does not contain four such countries. For example, in the United States map, the five states surrounding Nevada require three colours and Nevada itself then requires a fourth (see Figure 1.4.)

(iv) We cannot replace the word 'four' by 'three' since many maps (such as that in Figure 1.4) require four colours.

(v) Up to now we have not coloured the exterior (infinite) region surrounding the countries of the map, but for reasons that will shortly become clear, we shall generally choose to do so (as in Figure 1.5).

Problem 1.1

State how many colours are needed to colour each of the following maps, if:

(a) we do not colour the exterior region;

(b) we do colour the exterior region.

Give a corresponding colouring in each case.

Figure 1.4

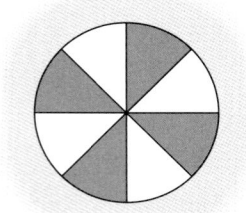

Figure 1.5

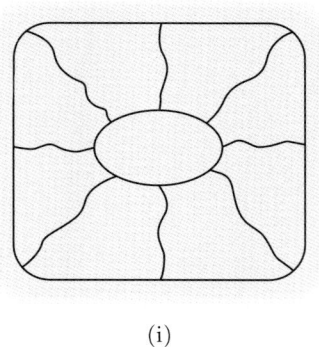

(i)

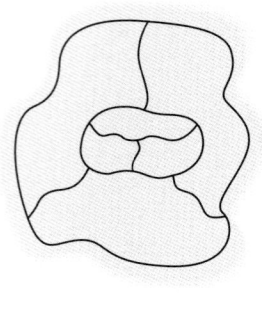

(ii)

Figure 1.6

If we have a four-colouring of the non-exterior regions of a map, then it may be necessary to find a different colouring if we wish also to colour the exterior region. For example, the map of Britain in Figure 1.1 has all four colours present around the coastline, so we apparently need a fifth colour for the sea. However, the Four-colour Theorem implies that we can re-colour this map so that only three of the colours appear around the coastline, thus allowing the sea to take the fourth colour.

1.2 Colouring maps on surfaces

Colouring maps on a plane is equivalent to colouring maps on the surface of a sphere. To see this, we use *stereographic projection*, as shown in Figure 1.7. As long as we are careful to arrange that the 'North Pole' N lies inside one of the countries of the map, we can project any map drawn on the sphere down to the plane, and we can project any map drawn in the plane up to the sphere, and the number of colours is not changed by such projections. We can thus treat maps on a plane and maps on a sphere as equivalent. We shall often refer to both as *spherical maps*.

This mapping is a bijection between the set of points of the sphere (other than N) and the set of points of the plane.

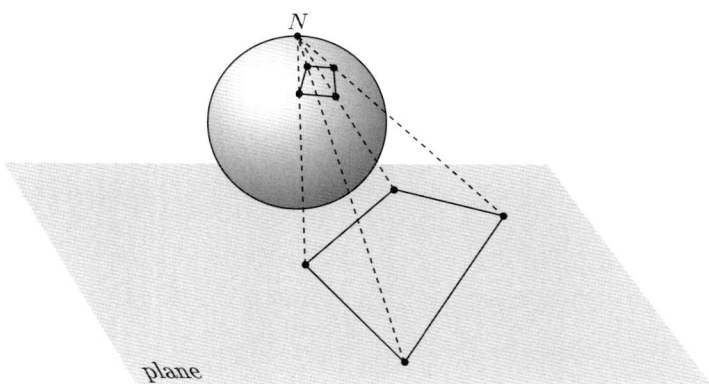

Figure 1.7

Note that the exterior region of the plane corresponds to the country on the sphere containing N. When we colour maps on a sphere this country can be regarded as being no different from any other country. Thus, from now on, when we colour maps on a plane, we colour all countries including the exterior one.

We now state the celebrated *Four-colour Theorem*.

> ### Theorem 1.1 Four-colour Theorem for spherical maps
> The countries of every spherical map can be coloured with at most four colours in such a way that any two countries with a common boundary line are assigned different colours.

An outline of a proof of the Four-Colour Theorem appears in Section 5.

We may also wish to consider other surfaces and ask, for example, how many colours are need to colour maps drawn on a torus. Here the 'magic number' is seven — every map drawn on a torus can be coloured with at most seven colours. Figure 1.8 shows a toroidal map that requires seven colours.

We prove this in Subsection 4.1.

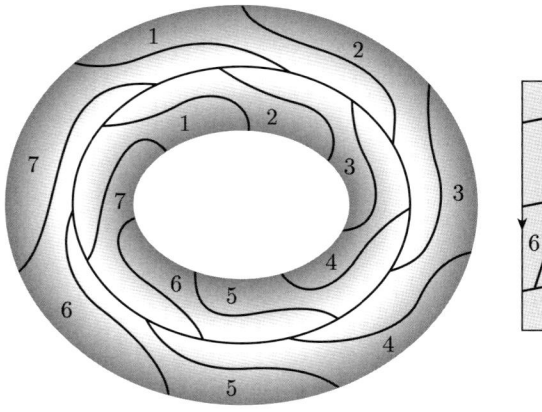 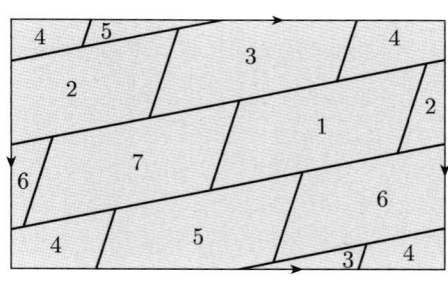

Figure 1.8

The torus appears here both in solid form and as a polygon with edges identified.

Similarly, we may ask how many colours are needed to colour maps drawn on a projective plane. Here the 'magic number' is six — every map drawn on a projective plane can be coloured with at most six colours. Figure 1.9 shows a map on the projective plane that requires six colours.

We prove this in Subsection 4.1.

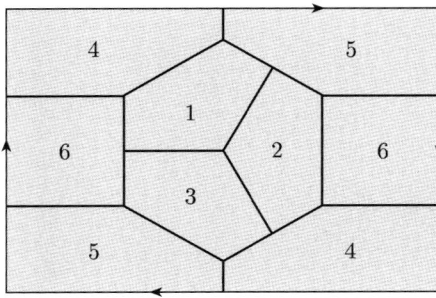

Figure 1.9

Before we can discuss the colouring of maps formally, we need to know precisely what is meant by a *map*. In fact, we shall impose some restrictions that do not significantly change the colouring problem but make it easier to investigate.

- We require each country to be homeomorphic to a disc — in particular, it must be in one piece and contain no holes (see Figure 1.10(a)).
- Since the colours on each side of a boundary line must be different, we do not allow maps in which a country meets itself (see Figure 1.10(b)).
- We assume that *each point where boundary lines meet lies on the boundary of at least three countries*, since meeting points lying on the boundaries of just two countries do not affect the colouring of the countries (see Figure 1.10(c)).

The projective plane appears here as a polygon with edges identified.

In many maps, each meeting point lies on the boundary of *exactly* three countries.

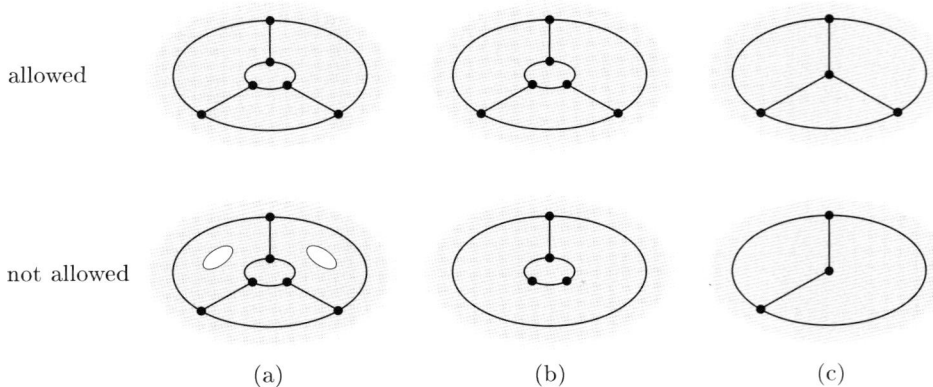

Figure 1.10

- We also assume that the map contains no countries with just one or two boundary lines, since any four-colouring of the countries in the rest of the map can easily be extended to these countries as well (see Figure 1.11).

A similar comment applies to countries with three boundary lines, but (for reasons that will become clear in Section 2) we choose to allow these.

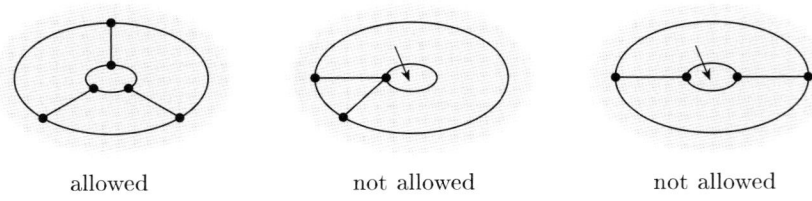

Figure 1.11

- Finally, we assume that no two countries have *more than one* boundary line in common — so we disallow the situation in Figure 1.12, in which the countries A and B have two edges in common.

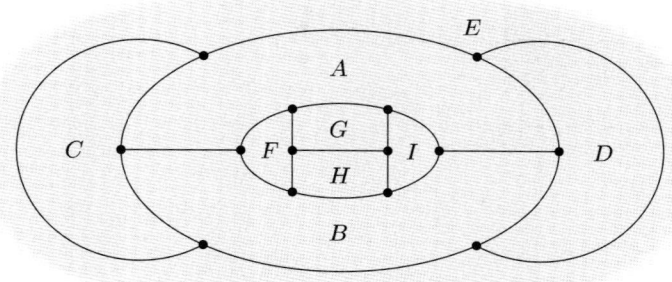

Figure 1.12

If we take the countries of a map to be its faces, the boundary lines to be its edges, and the meeting points of the boundary lines to be its vertices, the above restrictions imply that the map is a *subdivision* of the surface on which it is drawn, with some extra properties, as specified in the following definition.

Unit B1, Subsection 3.2.

Definition

A **map** M on a compact surface S is a finite subdivision of S with the following properties:
(a) no face meets itself at an edge;
(b) at least three edges meet at each vertex;
(c) each face has at least three edges;
(d) two faces can have at most one edge in common.

From now on, when discussing maps, we usually use the terms *face*, *edge* and *vertex* for country, boundary line and meeting point.

Remarks

(i) A *finite subdivision* is one with a finite number of faces, edges and vertices.
(ii) Properties (c) and (d) allow us to deduce that each face meets at least three others.
(iii) Provided M has at least one vertex, we can deduce from (b), (c) and (d) that it has at least four faces and at least four vertices.

One particular type of map has an important role to play in the investigation of colouring problems:

Definition

A **cubic map** is a map in which exactly three edges meet at each vertex.

Problem 1.2

Which of the following are maps? Which are cubic maps?

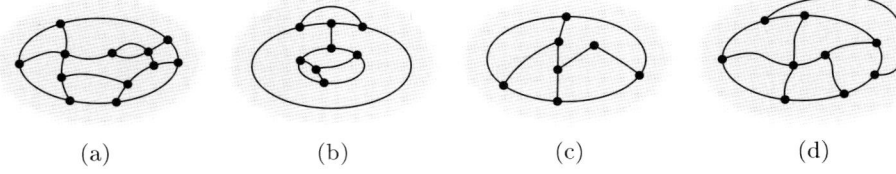

(a)   (b)   (c)   (d)

Figure 1.13

We conclude this subsection by returning to spherical maps and show that *when trying to prove the Four-colour Theorem, we may assume, whenever convenient, that all maps under consideration are cubic maps.*

This was first proved by the English mathematician Arthur Cayley in 1879.

To see why, suppose that at each vertex where more than three faces meet we stick a 'patch' over the vertex, as shown in Figure 1.14; then the resulting map is a cubic map. If we now colour the faces of this cubic map with four colours, we can obtain a four-colouring of the faces of the original map by shrinking each patch to a vertex.

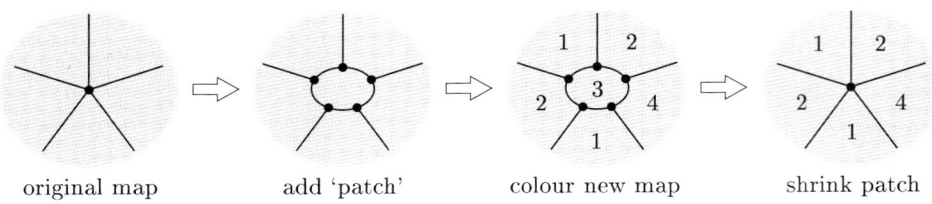

original map   add 'patch'   colour new map   shrink patch

Figure 1.14

Problem 1.3

What is wrong with the following argument?

> In the map on the right of Figure 1.14, only three colours appear around the vertex. Since we can carry out the same procedure at each vertex, we can therefore colour the entire map with just three colours.

1.3 Euler's Formula

A useful fact, which we shall prove in this subsection, is that:

> every spherical map has at least one face that meets five or fewer other faces.

Also the definition of a map tells us that each face must meet at least three others. Therefore a map must contain at least one of the types of face shown in Figure 1.15.

See Remark (ii) following the definition of a map.

 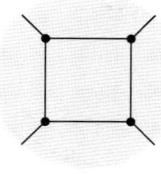 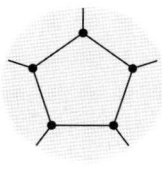

triangle   quadrilateral   pentagon

We refer to a face with three, four or five edges as a *triangle*, *quadrilateral* or *pentagon*, respectively.

Figure 1.15

To prove the quoted result requires some preliminary work.

Now, every map is a subdivision. Recall that *Euler's Formula* for a subdivision with V vertices, E edges and F faces on a compact surface with Euler characteristic χ is

$V - E + F = \chi$.

Unit B1, Subsection 3.3.

Every spherical map is a subdivision of the sphere, whose Euler characteristic is $\chi = 2$. We thus have the following result.

Unit B1, Subsection 3.3.

Theorem 1.2 Euler's Formula for spherical maps

For any spherical map with V vertices, E edges and F faces,

$V - E + F = 2$.

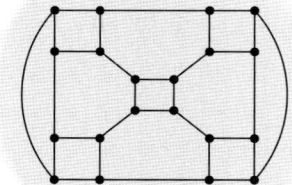

For example, for the spherical map in Figure 1.16,

$V = 20, E = 30, F = 12$, and $12 - 30 + 20 = 2$.

Figure 1.16

Problem 1.4

Verify Euler's Formula for each of the spherical maps in Figure 1.17.

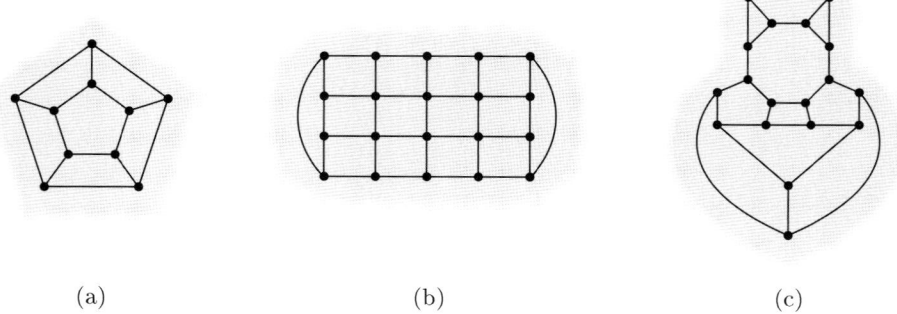

(a)   (b)   (c)

Figure 1.17

A related result concerns the numbers of faces of each type in a spherical map. We call it the *Face-counting Theorem*; for simplicity, we restrict our attention to cubic maps, so that exactly three edges meet at each vertex.

> **Theorem 1.3 Face-counting Theorem**
>
> Let M be a cubic spherical map, and for each positive integer k let F_k be the number of faces of M with k edges. Then
>
> $$3F_3 + 2F_4 + F_5 - F_7 - 2F_8 - 3F_9 - \cdots = \sum_{k \geq 3}(6-k)F_k = 12. \quad (1.1)$$

Note that there is no term involving F_6 and that, for any given map, this is a finite sum.

Before we prove this theorem, we illustrate it with the spherical map in Figure 1.16: the only non-zero terms are $F_4 = 7$, $F_6 = 4$ and $F_8 = 1$ (the exterior face), and the left-hand side of (1.1) becomes

$$0 + 14 + 0 - 0 - 2 - 0 - \cdots = 12.$$

Problem 1.5

Verify the statement of Theorem 1.3 for the cubic spherical map (c) in Problem 1.4.

Proof of Theorem 1.3 Suppose that the map has F faces, E edges and V vertices.

Counting the faces of each type, we obtain:

$$F = F_3 + F_4 + F_5 + F_6 + F_7 + F_8 + F_9 + \cdots = \sum_{k \geq 3} F_k. \quad (1.2)$$

We next count the edges around all the faces. Adding these together we obtain *twice* the total number of edges: each edge belongs to two faces and so is counted twice. We obtain:

$$2E = 3F_3 + 4F_4 + 5F_5 + 6F_6 + 7F_7 + 8F_8 + 9F_9 + \cdots = \sum_{k \geq 3} kF_k. \quad (1.3)$$

Finally, we count the vertices around each face. Each face with k edges has k vertices. Since the map is cubic, on adding these together we obtain *three times* the total number of vertices: each vertex belongs to three faces and is counted three times. We obtain:

$$3V = 3F_3 + 4F_4 + 5F_5 + 6F_6 + 7F_7 + 8F_8 + 9F_9 + \cdots = \sum_{k \geq 3} kF_k. \quad (1.4)$$

Substituting (1.2), (1.3) and (1.4) into Euler's Formula $V - E + F = 2$, or equivalently $12 = 6V - 6E + 6F$, we obtain:

Theorem 1.2.

$$12 = 2(3V) - 3(2E) + 6F$$
$$= 2\sum_{k \geq 3} kF_k - 3\sum_{k \geq 3} kF_k + 6\sum_{k \geq 3} F_k = \sum_{k \geq 3}(6-k)F_k. \quad \blacksquare$$

Problem 1.6

Use Theorem 1.3 to deduce that:

(a) every cubic spherical map containing only triangles and hexagons has exactly four triangles;

(b) every cubic spherical map containing only pentagons and hexagons has exactly twelve pentagons.

A *hexagon* is a face with six edges.

It follows from Theorem 1.3 that every spherical map has at least one face with five or fewer edges — that is, a triangle, a quadrilateral or a pentagon. This is because if a cubic map had no such faces, then $F_3 = F_4 = F_5 = 0$, and the left-hand-side of (1.1) would be negative or zero while the right-hand side is positive ($= 12$), giving a contradiction.

We now generalize this result by proving that every spherical map (not necessarily cubic) has at least one face that meets five or fewer other faces. We first derive a simple lemma that will be used throughout this unit; it applies to maps drawn on any surface.

Lemma 1.4
(a) In any map with V vertices and E edges, $V \leq \frac{2}{3}E$.
(b) In any map with F faces and E edges, $F \leq \frac{2}{3}E$.

Proof
(a) By the definition of a map, at least three edges meet at each of the V vertices. But each edge has two vertices. So, counting edges, we obtain $2E \geq 3V$. Thus, $E \geq \frac{3}{2}V$ and $V \leq \frac{2}{3}E$.

(b) By the definition of a map, each of the F faces has at least three edges. But each edge belongs to two faces. So, counting edges, we obtain $2E \geq 3F$. Thus, $E \geq \frac{3}{2}F$ and $F \leq \frac{2}{3}E$. ∎

We also note that the *average* number of edges per face is $2E/F$. This is because the total number of 'edge-counts' around the faces is $2E$: each edge belongs to two faces and is counted twice.

We can now prove the promised theorem.

Theorem 1.5
Every spherical map has at least one face that meets five or fewer other faces.

Proof Suppose that such a map has V vertices, E edges and F faces.

Substituting the inequality $V \leq \frac{2}{3}E$ from Lemma 1.4(a) into Euler's Formula for spherical maps, we obtain

$$2 = V - E + F \leq \tfrac{2}{3}E - E + F,$$

which can be rearranged to give $E \leq 3F - 6$.

Hence, on multiplying by $2/F$, we obtain

$$\frac{2E}{F} \leq \frac{2(3F-6)}{F} = 6 - \frac{12}{F} < 6.$$

Since $2E/F$ (the *average* number of edges per face) is less than 6, it follows that *at least one* face must have fewer than six edges. Thus, at least one face must meet five or fewer other faces. ∎

Problem 1.7

Give an example of a spherical map in which every face meets *at least* five other faces.

1.4 The Five-colour Theorem

In this subsection we prove that the faces of every spherical map can be coloured with five colours in such a way that neighbouring faces are coloured differently. The proof is by mathematical induction on the number of faces, and uses Theorem 1.5: every map has a face that meets at most five other faces.

In order to illustrate the underlying ideas, we first prove the following weaker result.

> **Theorem 1.6 Six-colour Theorem for spherical maps**
> The faces of every spherical map can be coloured with at most six colours in such a way that any two faces with a common edge are assigned different colours.

Proof The proof is by mathematical induction on the number of faces.

The result is clearly true for spherical maps with up to six faces (since all the faces can be given different colours).

We now assume that

   all spherical maps with k faces can be coloured with at most six colours,

and show that

   all spherical maps with k + 1 faces can be so coloured.

This is our induction hypothesis.

Let M be a spherical map with $k + 1$ faces. By Theorem 1.5, M has a face A that meets at most five other faces. If we remove one of the edges e bounding A (say the edge separating face A from face B), then the resulting map N has k faces, one of which (let us call it C) replaces faces A and B (Figure 1.18). By our induction hypothesis, we can colour the faces of the map N with six colours in such a way that neighbouring faces are coloured differently.

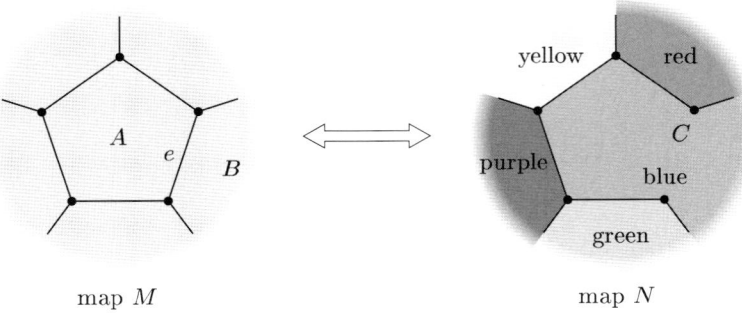

map M \qquad map N

Figure 1.18

The portion of N shown requires five of the six colours that we assume are used for the whole of N.

We now reinstate the edge e, and colour M as follows. For the faces other than A and B, we use the colours of the corresponding faces of N. For face B, we use the colour that was used on face C of N. Now the only face left to colour is A (Figure 1.19). Since A has at most five neighbours, and six colours are available, there is a spare colour that can be used. This gives a six-colouring of the faces of the original map M.

The result now follows by mathematical induction. ∎

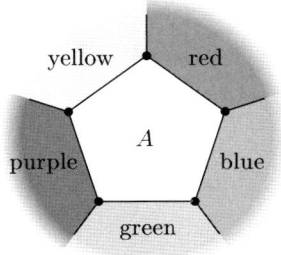

Figure 1.19

Proving that the faces of every map can be coloured with *five* colours requires a little more effort, although the basic approach is the same.

> **Theorem 1.7 Five-colour Theorem for spherical maps**
>
> The faces of every spherical map can be coloured with at most five colours in such a way that any two faces with a common edge are assigned different colours.

Proof The proof is by mathematical induction on the number of faces.

This proof is not assessed.

The result is clearly true for spherical maps with up to five faces (since all the faces can be given different colours).

We now assume that

all spherical maps with k faces can be coloured with at most five colours,

and show that

all spherical maps with $k+1$ faces can be so coloured.

This is our induction hypothesis.

Let M be a spherical map with $k+1$ faces. By Theorem 1.5, M has a face A that meets at most five other faces. If we remove one of the edges e bounding A (say the edge separating face A from face B), then the resulting map N has k faces, one of which (let us call it C) replaces faces A and B. By our induction hypothesis, we can colour the faces of the map N with five colours in such a way that neighbouring faces are coloured differently.

We now reinstate the edge e, and colour M as follows. For the faces other than A and B, we use the colours of the corresponding faces of N. For face B, we use the colour that was used on face C of N. Now the only face left to colour is A. Since A has at most five neighbours and five colours are available, there is a spare colour that can be used to colour A, *unless A is bounded by five faces of different colours*, as illustrated in Figure 1.20; in this case, there is no spare colour that can be used to colour A.

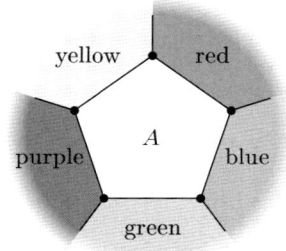

Figure 1.20

We have chosen five colours surrounding A in a particular order; clearly any five colours in any order would give a similar argument.

To overcome this difficulty, we consider just (say) the red and green neighbours of A, and investigate whether there is a path of red and green faces joining them.

Two situations can arise, as Figure 1.21 illustrates.

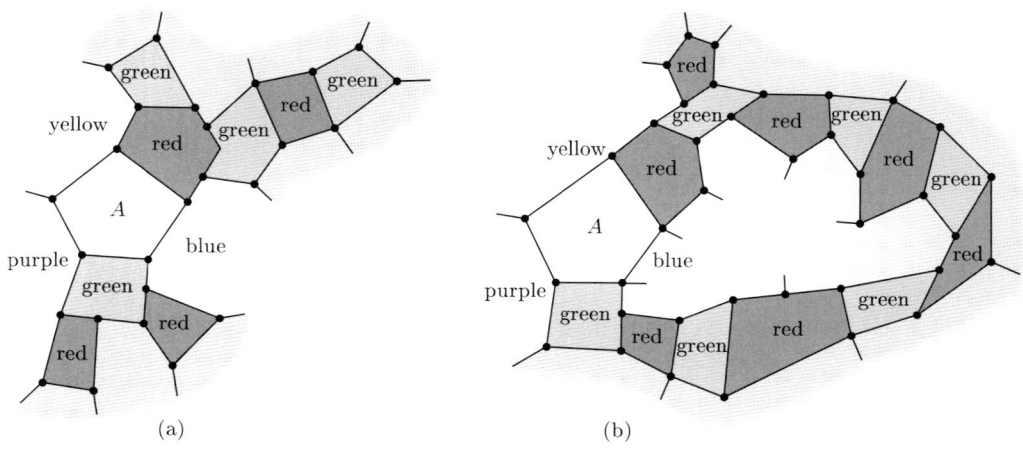

Figure 1.21

In case (a), all the red and green faces reachable from the red neighbour of A are different from those reachable from the green one, and so there is no such red–green path of faces. In this case, we can interchange the colours in the red–green part at the top, without affecting the colouring of the rest of the map, as shown in Figure 1.22. This replaces the red neighbour of A by a green one, so that A can be coloured red. This completes the five-colouring of the original map M in case (a).

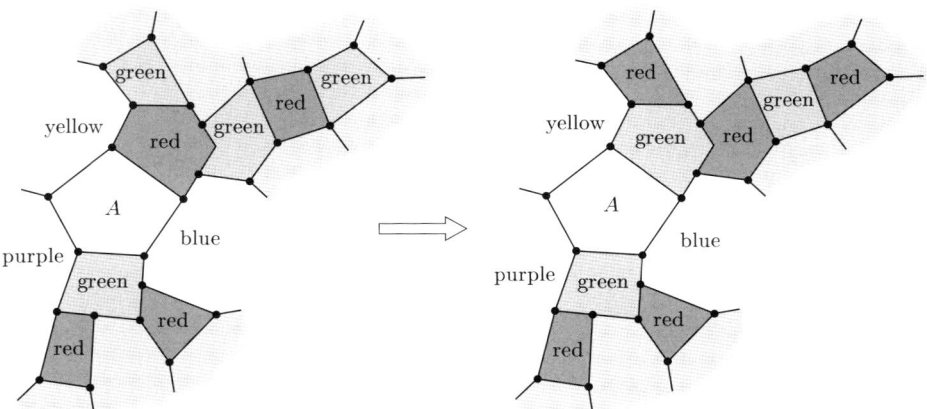

Figure 1.22

In case (b), the red and green neighbours of A are joined by a red–green path of faces, and interchanging the colours would not help us as the face A would still have a red neighbour and a green one. But in this case, there can be no blue–yellow path of faces between the blue and yellow neighbours of A, because the red–green path 'gets in the way'. We can therefore interchange the colours in the blue–yellow part on the right-hand side without affecting the colouring of the rest of the map, as shown in Figure 1.23. This replaces the blue neighbour of A by a yellow one, so that A can be coloured blue. This completes the five-colouring (as above) of the original map M in case (b).

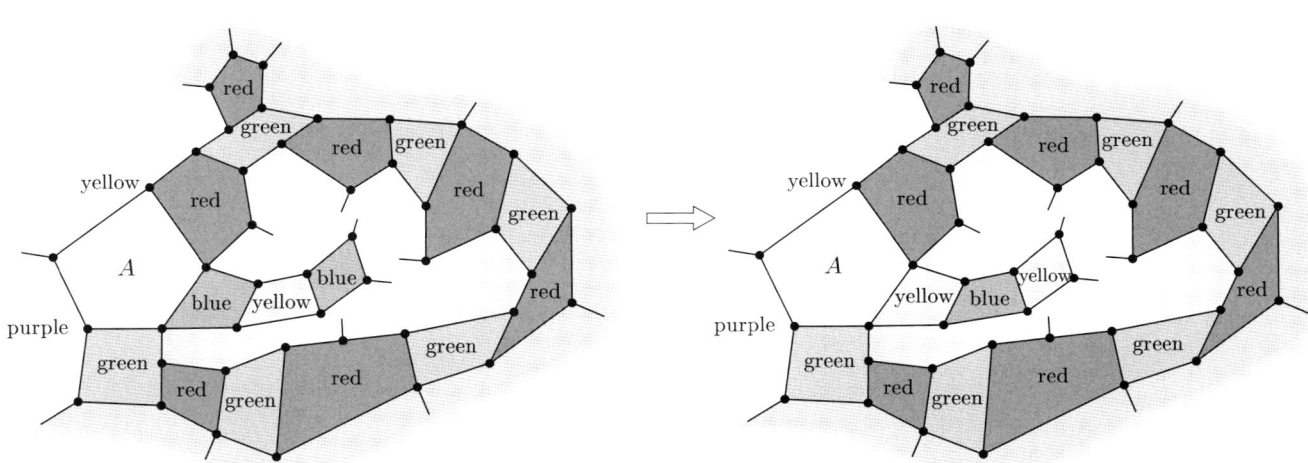

Figure 1.23

The result now follows by mathematical induction. ∎

In Section 5 we show how these ideas were used by Kempe in his incorrect proof of the Four-colour Theorem (Theorem 1.1), and we present a map constructed by Heawood that dramatically illustrates the flaw in Kempe's argument. The error in Kempe's argument was a major one, and many years elapsed before the Four-colour Theorem was proved. However, Kempe's attempted proof contained two fundamental ideas — an *unavoidable set* and a *reducible configuration* — that led eventually to the solution of the problem by Appel and Haken in the 1970s. In Section 5 we explain these ideas and use them to outline Appel and Haken's proof of the Four-colour Theorem.

2 Colouring graphs

> After working through this section, you should be able to:
> ▶ explain the terms *graph, subgraph, connected graph, planar graph, complete graph* and *complete bipartite graph*;
> ▶ explain the connection between maps and planar graphs;
> ▶ explain what are meant by a *k-colouring* of a graph and the *chromatic number* of a graph;
> ▶ deduce the *Five-colour Theorem* for planar graphs from the corresponding theorem for spherical maps.

In this section we present an alternative approach to map colouring, using the 'dual' concept of colouring the vertices of a graph drawn on the plane or the sphere. You have already met the idea of duality in *Unit B2*, but here we recast it in the context of maps.

Unit B2, Subsection 1.2.

Suppose that we have a map on the sphere, as shown in Figure 2.1(a). We now draw a small circle in each face, and join the circles in neighbouring faces by lines, as in Figure 2.1(b). The diagram of circles and lines, shown in Figure 2.1(c), is an example of a *planar graph*; it is a convenient device for representing the adjacencies between the faces of a map.

Recall that we include the exterior face.

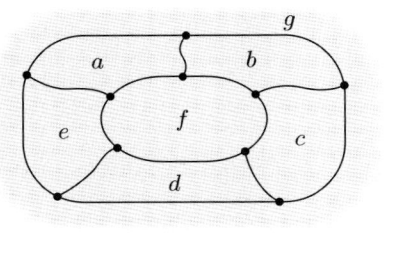
(a)
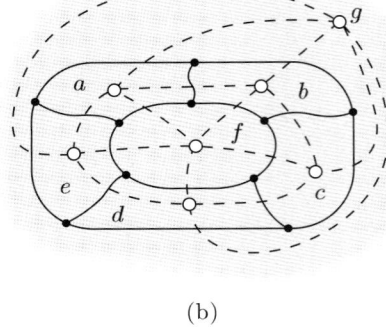
(b)
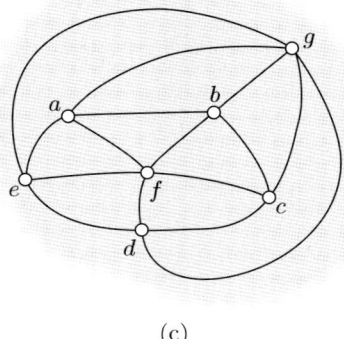
(c)

Figure 2.1

If we now colour the faces of the map so that faces sharing an edge are coloured differently, then we obtain a corresponding colouring of the points of the planar graph, as Figure 2.2 illustrates. This colouring of the points has the following property: whenever two points are joined by a line, they must be coloured differently.

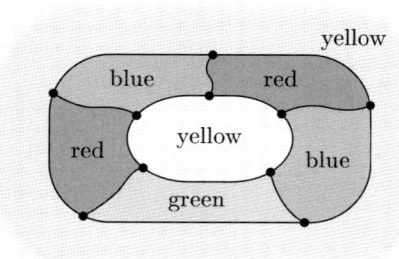
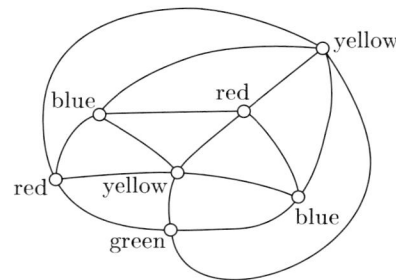

Figure 2.2

In this section we investigate graphs in general. We find that many of the results in Section 1 concerning maps and their colourings give rise to analogous results concerning planar graphs and their colourings. We explore these ideas further in Sections 3 and 4.

2.1 Graphs

Before investigating planar graphs, we introduce the idea of a graph in general.

> ### Definition
> A **graph** is a diagram consisting of a non-empty set of points, called **vertices**, and a (possibly empty) set of elements called **edges**. Each edge joins two distinct vertices.
>
> Two vertices are **adjacent** if there is an edge joining them.

Remarks

(i) Since each edge joins distinct vertices, we do not allow an edge to join a vertex to itself (sometimes called a *loop*).
(ii) We do not allow two or more edges joining the same pair of vertices (sometimes called *multiple edges*).
(iii) The edge joining two vertices a and b can be written as ab or ba.

The type of graph defined here, without loops or multiple edges, is sometimes called a *simple graph*.

Four graphs are shown in Figure 2.3.

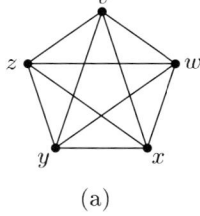
(a)
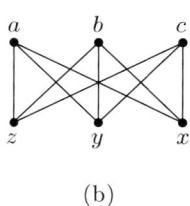
(b)
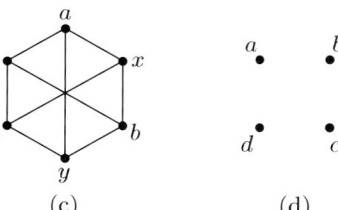
(c)   (d)

Notice that edges may cross at points that are not vertices of the graph: for example, the centre of graph (c) is not a vertex.

Figure 2.3

In graph (a) there are five vertices $\{v, w, x, y, z\}$ and ten edges that join these vertices in pairs:

$$\{vw, vx, vy, vz, wx, wy, wz, xy, xz, yz\};$$

here, every two vertices are adjacent.

In graphs (b) and (c) there are six vertices $\{a, b, c, x, y, z\}$ and nine edges:

$\{ax, ay, az, bx, by, bz, cx, cy, cz\}$.

Here, each of the vertices a, b, c is adjacent to each of the vertices x, y, z, but not to each other.

In graph (d) there are four vertices and no edges.

In general, a graph can be drawn in many different ways: for example, diagrams (b) and (c) are two drawings of the same graph, since they have the same sets of vertices and edges.

Problem 2.1

Draw the graphs with the following sets of vertices and edges:

(a) vertices $\{u, v, w, x\}$, edges $\{uv, ux, uw, vw, wx\}$;
(b) vertices $\{a, b, x, y, z\}$, edges $\{ax, ay, az, bx, by, bz\}$.

In each graph, to which vertices is vertex x adjacent?

Two important types of graph are the *complete graphs* and the *complete bipartite graphs*. These play an important role throughout this unit.

Definition

The **complete graph** K_n has n vertices, each of which is joined to each other vertex by exactly one edge.

The **complete bipartite graph** $K_{r,s}$ has two subsets of vertices, containing r and s vertices respectively; each of the r vertices is joined to each of the s vertices by exactly one edge, and there are no other edges.

For example, in Figure 2.3 diagram (a) is the complete graph K_5 and diagrams (b) and (c) are two drawings of the complete bipartite graph $K_{3,3}$.

The definition of a graph does not require the vertices to be labelled. So, if the vertex labels are removed from Figure 2.3, we still have a representation of K_5 in (a) and of $K_{3,3}$ in (b) and (c).

Problem 2.2

(a) Draw the complete graphs K_4 and K_6.
(b) How many edges has the complete graph K_n?

Problem 2.3

(a) Draw the complete bipartite graphs $K_{2,4}$ and $K_{1,5}$.
(b) How many edges has the complete bipartite graph $K_{r,s}$?

It is also useful to have a term for the number of edges meeting at a vertex.

Definition

In a graph, the **degree** of a vertex v is the number of edges meeting at v.

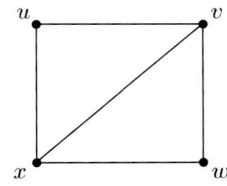

Figure 2.4

For example, in Figure 2.4, each of the vertices u and w has degree 2 and each of the vertices v and x has degree 3.

Problem 2.4

What are the degrees of the vertices in K_n and in $K_{r,s}$?

Problem 2.5

What is the relationship between the sum of the degrees of all the vertices in a graph and the number of edges in the graph?

We also need the concept of a *subgraph* of a graph.

> **Definition**
>
> A **subgraph** of a graph G is a graph all of whose vertices are vertices of G and all of whose edges are edges of G.

For example, Figure 2.5 shows three subgraphs of the complete graph K_5.

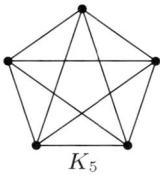
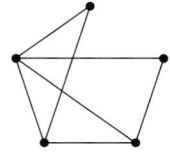
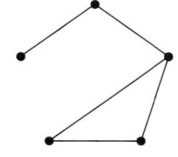

Figure 2.5

Problem 2.6

Which of the graphs in Figure 2.6 are subgraphs of the graph in Figure 2.7?

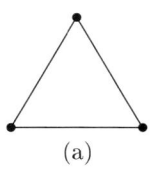
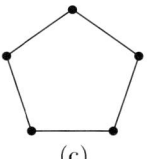
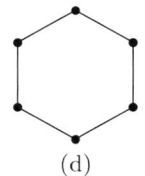
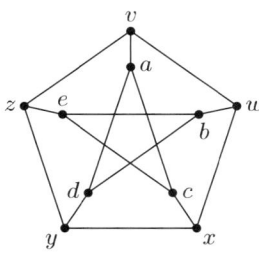

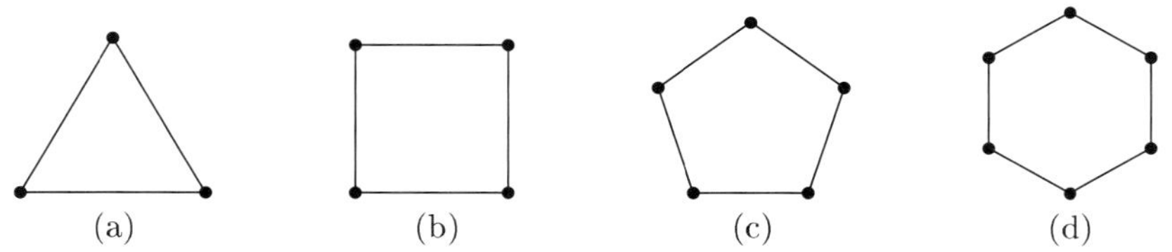

Figure 2.6

Figure 2.7

We assume throughout this section that our graphs are *connected* — that is, each is in one piece: this means that there is a path of edges between any two vertices. For example, Figure 2.8 illustrates a connected graph and a disconnected one: in the connected graph there is a path xyz (consisting of the edges xy and yz) between the vertices x and z, whereas in the disconnected graph there is no path between these vertices.

In fact, there are several such paths.

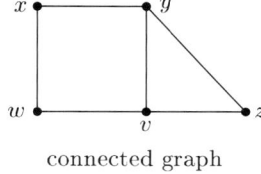

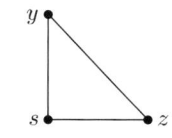

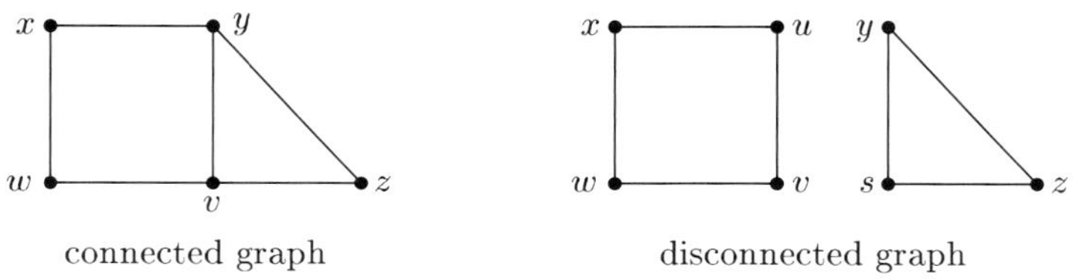

connected graph   disconnected graph

Figure 2.8

2.2 Planar graphs

Consider now the drawing of the complete graph K_4 in Figure 2.9, with four vertices $\{a, b, c, d\}$ and six edges $\{ab, ac, ad, bc, bd, cd\}$. In drawing (a) the edges ac and bd cross at a point that is not a vertex, and we may ask whether this graph can be redrawn in the plane without any crossings. Two possibilities are shown in drawings (b) and (c).

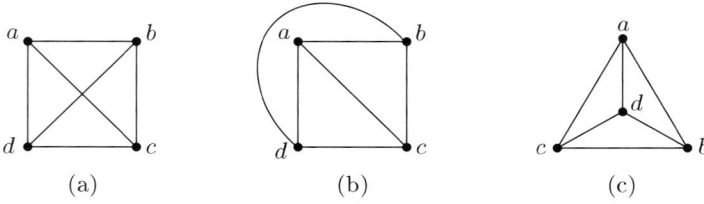

Figure 2.9   The complete graph K_4.

With this example in mind, we make the following definition.

> **Definition**
>
> A graph G is **planar** if it can be redrawn in the plane or on the sphere in such a way that no two edges meet except at a vertex. Any such drawing is a **plane drawing** of G.

We use the usual term *planar graph*, although we are mainly interested in the equivalent problem of drawing graphs on the sphere.

Thus, the complete graph K_4 is a planar graph, since it can be redrawn without crossings. Similarly, the graph in Figure 2.10 is planar, because it can be 'unravelled' as shown.

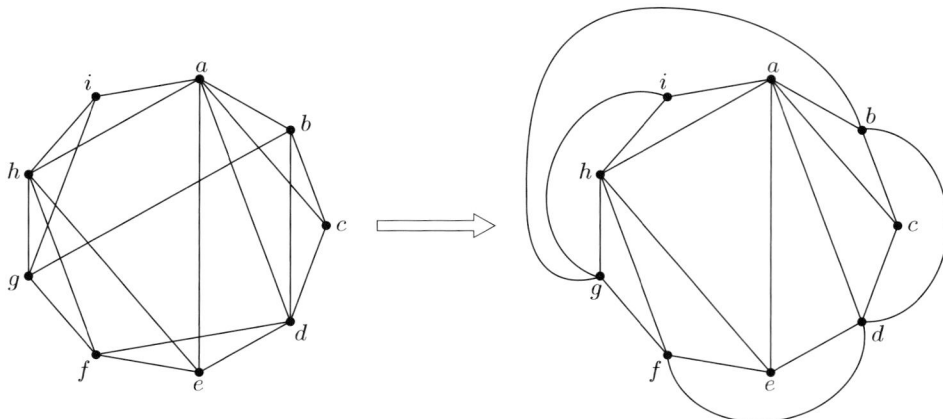

Figure 2.10

Problem 2.7

Show that the graphs in Figure 2.11 are planar, by finding a plane drawing of each.

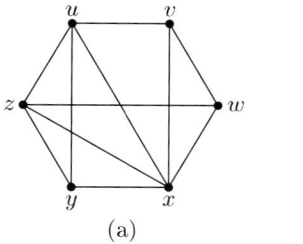 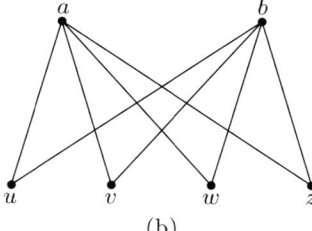

(a) (b)

Figure 2.11

Not all graphs are planar. For example, every attempt to draw the complete bipartite graph $K_{3,3}$ in the plane yields a drawing with at least one crossing. To see why this is, notice from Figure 2.12(a) that the graph has a cycle of six vertices ($axbycza$), each adjacent to the previous vertex and returning to the first one. If there were a plane drawing of $K_{3,3}$ containing no crossings, then this cycle would have to appear as a hexagon, not necessarily regular, as shown in Figure 2.12(b).

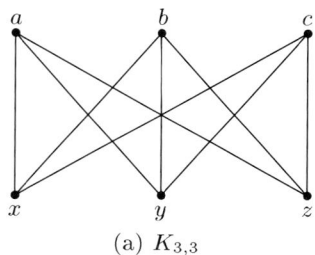 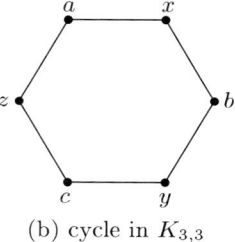

(a) $K_{3,3}$ (b) cycle in $K_{3,3}$

Figure 2.12

We must now insert the edges ay, bz and cx. Only one of them can be drawn *inside* the hexagon, since any two would cross as in Figure 2.13(a). Similarly, only one of them can be drawn *outside* the hexagon, since any two would cross as in Figure 2.13(b).

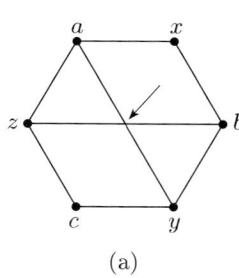 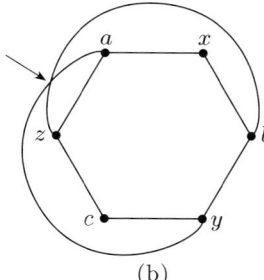

(a) (b)

Figure 2.13

It is therefore impossible to insert all three of these edges without creating a crossing. This demonstrates that $K_{3,3}$ is non-planar.

Problem 2.8

Give a similar argument to demonstrate that the complete graph K_5 is non-planar.

Note that any *subgraph* H of a planar graph G is planar: we can draw G in the plane without crossings, and removing the vertices and edges not in H gives a plane drawing of H. For example, the complete graph K_4 and the complete bipartite graph $K_{2,3}$ are planar, and thus any subgraph of either of them is planar (see Figure 2.14).

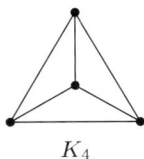 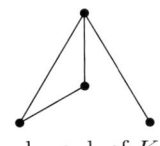 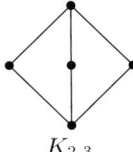 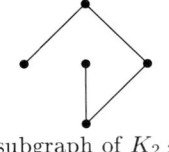

K_4 subgraph of K_4 $K_{2,3}$ subgraph of $K_{2,3}$

Figure 2.14

Similarly, any graph G that *contains* a non-planar subgraph H is non-planar: if G could be drawn in the plane without crossings, then so could H, contradicting the fact that H is non-planar. For example, the complete graph K_5 and the complete bipartite graph $K_{3,3}$ are non-planar, and thus any graph that contains either of them is non-planar (see Figure 2.15).

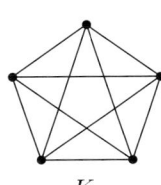 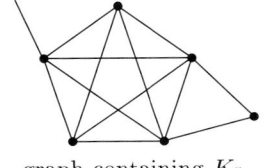 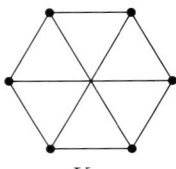 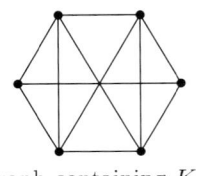

K_5 graph containing K_5 $K_{3,3}$ graph containing $K_{3,3}$

Figure 2.15

Problem 2.9

(a) For which values of n is the complete graph K_n planar?

(b) For which values of s are the complete bipartite graphs $K_{1,s}$ and $K_{2,s}$ planar?

(c) For which values of r and s is the complete bipartite graph $K_{r,s}$ planar?

We conclude this subsection by noting that a plane drawing of a connected planar graph on the sphere defines a subdivision of the sphere, with the vertices and edges of the graph corresponding to the vertices and edges of the subdivision. The connectedness of the subdivision is ensured by the connectedness of the graph. The planarity of the graph ensures that no two edges of the subdivision have internal points in common. This equivalence proves useful in the next subsection.

2.3 Colouring planar graphs

At the beginning of this section we showed how a colouring of the faces of a map leads naturally to a colouring of the vertices of a related planar graph. In this colouring, any two vertices joined by an edge must have different colours. More generally, we can try to colour the vertices of any graph (not necessarily planar) in such a way that adjacent vertices are coloured differently.

> **Definition**
>
> A **k-colouring** of a graph G is an assignment of up to k colours to the vertices of G in such a way that each vertex is assigned just one colour and adjacent vertices are assigned different colours. If G has a k-colouring, then G is **k-colourable**.
>
> The **chromatic number** of G, denoted by $\text{ch}(G)$, is the smallest number k for which G is k-colourable.

Most books use the symbol $\chi(G)$ for the chromatic number of G; we have not done so due to possible confusion with the Euler characteristic.

For example, the graph G in Figure 2.16 is four-colourable, as shown in (a), and is also three-colourable, as shown in (b). It is not, however, two-colourable, since it contains a triangle. Thus, its chromatic number $\text{ch}(G)$ is 3.

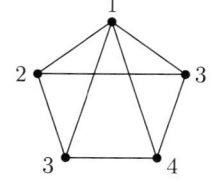
(a) G is four-colourable

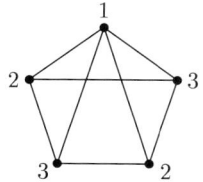
(b) G is three-colourable

For convenience, we use numbers for the colours.

Figure 2.16

Problem 2.10

Determine $\text{ch}(G)$ for each of the graphs G in Figure 2.17.

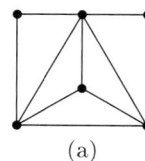
(a)

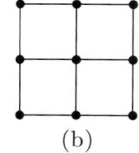
(b)

Figure 2.17

Problem 2.11

Write down the chromatic number of each of the following graphs:
(a) the complete graph K_n;
(b) the complete bipartite graph $K_{r,s}$;
(c) the graph C_n, consisting of a single cycle with n vertices and n edges.

The graphs C_3, C_4, C_5 and C_6 appeared in Figure 2.6.

There are graphs with arbitrarily large chromatic number: for example, the complete graph K_n has chromatic number n, for any value of n. However, as we shall see, the chromatic number of any *planar* graph is at most 4.

Duality

The connection between colouring the faces of a map and colouring the vertices of a planar graph becomes clearer when we recall the idea of duality, introduced in *Unit B2* and referred to briefly at the beginning of this section.

Unit B2, Subsection 1.2.

In *Unit B2* we saw that if S is any subdivision of a compact surface then we can form another subdivision S^*, called a *dual subdivision*, which is constructed as follows (see Figure 2.18):

(a) place a new vertex inside each face of the subdivision S;

(b) join two new vertices by a new edge if and only if they lie in faces of S with an edge in common.

Each new edge should be drawn so that it crosses the original edge.

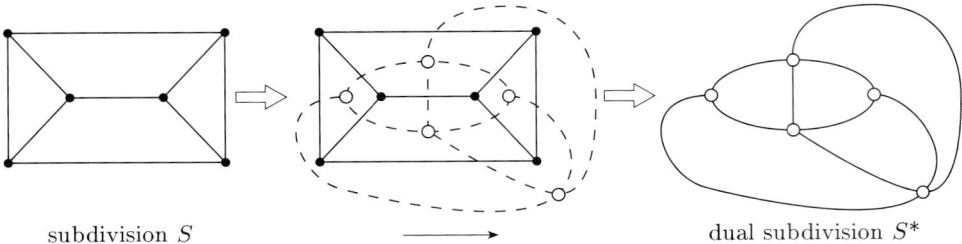

subdivision S ⟶ dual subdivision S^*

Figure 2.18

Since, by definition, a map M is a subdivision, we can create its dual subdivision M^*. By definition of a map, no face meets itself at an edge, so M^* has no loops. Also, no two faces of M have more than one edge in common, so M^* has no multiple edges. Thus, the dual subdivision M^* is a graph with vertices and edges corresponding to those of the dual subdivision. Furthermore, from the definition of a subdivision, M has no crossings. Thus, from the way subdivisions are constructed, neither does M^*. We deduce that, if M is a spherical map, then M^* is a plane drawing of a planar graph.

We saw at the end of Subsection 2.2 that a plane drawing of a connected planar graph G defines a subdivision of the sphere, and so we can create its dual subdivision G^*. We now demonstrate that, if each vertex of G has degree at least 3, then G^* is a map. Since G has no loops or multiple edges, no face of G^* meets itself and 2 faces have at most one edge in common. Furthermore, each face of G has at least 3 edges, and so at least 3 edges meet at each vertex of G^*. Finally, each vertex of G has degree at least 3, so each face of G^* has at least 3 edges. This relationship will prove useful shortly.

We can now use these duality ideas to investigate the colouring of subdivisions and planar graphs. Our first result applies to subdivisions of any compact surface.

Theorem 2.1

Let S be a subdivision of a compact surface and let S^* be a dual subdivision.

The faces of S can be coloured with k colours so that neighbouring faces are coloured differently if and only if the vertices of S^* can be coloured with k colours so that adjacent vertices are coloured differently.

By *neighbouring* faces we mean faces with a common edge.

Proof Assume first that the faces of S can be coloured with k colours so that neighbouring faces are coloured differently. We now give each vertex of S^* the same colour as the face containing it: this produces a k-colouring of the vertices of S^*. No two adjacent vertices of S^* can be coloured the same, since otherwise the faces of S containing them would be neighbours and have the same colour, which is not permitted. Thus, the vertices of S^* can be coloured with k colours so that adjacent vertices are coloured differently.

The converse argument is similar. ■

Problem 2.12

Prove the converse part of Theorem 2.1.

As a consequence of Theorem 2.1 we can deduce the following result from Theorem 1.7, the Five-colour Theorem for spherical maps.

> ### Theorem 2.2 Five-colour Theorem for planar graphs
> Every planar graph is five-colourable.

Proof If the planar graph is disconnected, then we can colour each connected part independently, so we can assume that our planar graph G is connected. We make a plane drawing of this connected planar graph, which we can consider to be a subdivision of the sphere.

We saw this at the end of Subsection 2.2.

Any vertex of G of degree less than 3 (indeed of degree less than 5) can be coloured with one of the five colours after all other vertices have been coloured, in such a way that its colour is different from its adjacent vertices. So we may assume that each vertex of G has degree at least 3. In these circumstances, as we saw above, the dual G^* of G is a spherical map.

By Theorem 1.7, we can colour the faces of G^* with five colours so that neighbouring faces are coloured differently. Thus, by Theorem 2.1 with $k = 5$, we can colour the vertices of G with five colours so that adjacent vertices are assigned different colours. ■

In the same way, once we have proved the Four-colour Theorem for spherical maps (Theorem 1.1), we can deduce the Four-colour Theorem for planar graphs.

> ### Theorem 2.3 Four-colour Theorem for planar graphs
> Every planar graph is four-colourable.

In Section 4 we use the idea of duality to relate the colouring of the vertices of graphs to the colouring of the faces of a map for other compact surfaces.

3 Embedding graphs on surfaces

> After working through this section, you should be able to:
> ▶ explain the terms *embedding* and *simple embedding*;
> ▶ define the *orientable genus* and *non-orientable genus* of a graph;
> ▶ state Euler's Formula for surfaces of a given genus;
> ▶ state the *Ringel–Youngs Theorem* for graphs embedded on an orientable or non-orientable surface.

In Section 2 we were concerned with drawing graphs on a sphere (or, equivalently, a plane), and we saw that the complete graph K_4 can be drawn without crossings on a sphere, whereas the complete graph K_5 cannot. Similarly, we saw that the complete bipartite graph $K_{2,3}$ can be drawn without crossings on a sphere, whereas the complete bipartite graph $K_{3,3}$ cannot.

If a graph can be drawn without crossings on a surface, we say that the graph *can be embedded*, or *has an embedding*, on the surface: for example, K_4 and $K_{2,3}$ can be embedded on a sphere, whereas K_5 and $K_{3,3}$ cannot. In this section we investigate what happens when we try to embed graphs on compact surfaces without boundary — these include *orientable* surfaces such as the torus and *non-orientable* surfaces such as the projective plane and Klein bottle. In each case we seek the 'simplest' compact surface without boundary on which a given graph can be embedded: for example, we shall see that the simplest orientable surface on which K_5 or $K_{3,3}$ can be embedded is the torus.

In this section we generally represent surfaces as polygons with edge identifications.

3.1 Revisiting surfaces

We start by considering the graph K_5 embedded on a torus; this can be done in several ways, as shown in Figure 3.1.

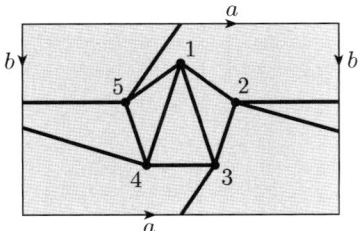 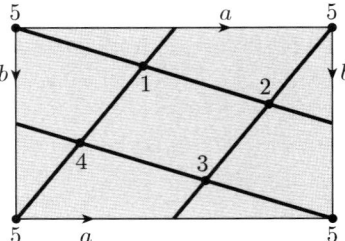 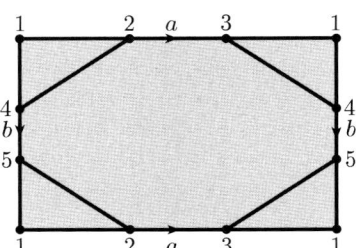

Figure 3.1

These embeddings are all examples of **simple embeddings**, in which *each face is homeomorphic to an open disc*. In general, a given graph may have both simple and non-simple embeddings on a surface. For example, consider the following drawings of the graph K_4 on a torus. In Figure 3.2(a), face 4 is not homeomorphic to an open disc, so this is not a simple embedding. However, if we redraw the graph as in Figure 3.2(b), then each face *is* homeomorphic to an open disc, so this is a simple embedding.

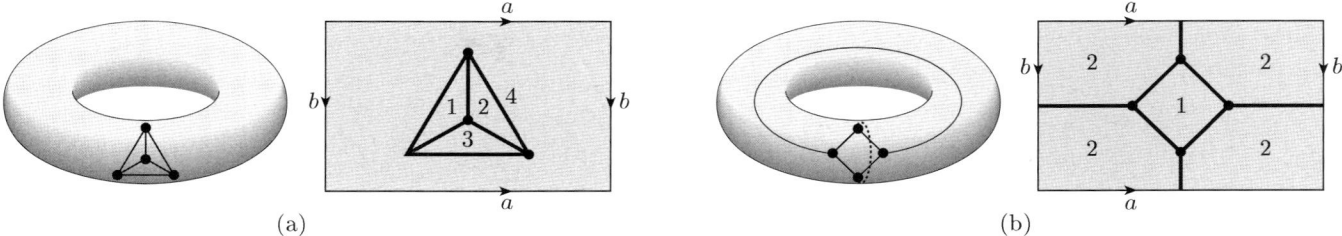

(a)   (b)

Figure 3.2

From now on, we always require that our drawings of graphs on surfaces are simple embeddings. Thus if we say that a graph can be embedded on a surface, we mean that a *simple* embedding of the graph on the surface exists.

Problem 3.1

Find a simple embedding of $K_{3,3}$ on a torus.

We deduced at the end of Subsection 2.2 that a connected planar graph on the sphere defines a subdivision of the sphere. In the same way, we can deduce that a simple embedding of a connected graph on any compact surface without boundary defines a subdivision of that surface. This fact will prove useful shortly.

Embeddings on orientable surfaces

Recall from *Unit B3* that orientable surfaces without boundary can be classified as follows.

Unit B3, Theorem 4.5.

> ### Theorem 3.1 Orientable surfaces
> Every compact orientable surface without boundary is topologically equivalent to a sphere with h handles, for some non-negative integer h.

When $h = 0$ we have a sphere; when $h = 1$ we have a torus; when $h = 2$ we have a two-fold torus; and for general h, we have an h-fold torus. Any compact orientable surface that is topologically equivalent to a sphere with h handles, or equivalently to an h-fold torus, is called an **orientable surface of genus h**, denoted by O_h (see Figure 3.3).

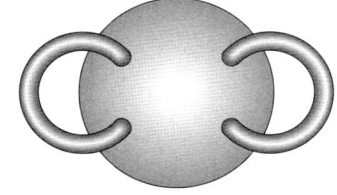

surface O_2

Figure 3.3

Given any graph, we can always find an orientable surface on which it can be embedded: we draw the graph on a sphere, using crossings as necessary, and then replace each crossing with a 'bridge' or handle, with one edge going over the bridge and one under (see Figure 3.4).

Although we do not prove this, such a use of handles always creates *simple* embeddings.

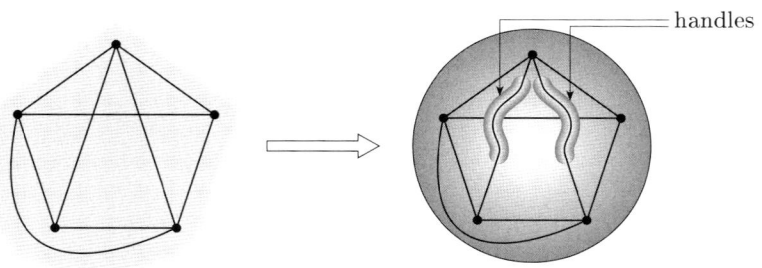

Figure 3.4

However, this method generally uses more handles than we need, and for each graph it is natural to ask for the *smallest* possible number of handles. Since the complete graph K_5 can be embedded on a torus, but not on a sphere, we can say that the torus is the 'simplest' orientable surface (that is, of smallest genus) on which K_5 can be embedded. More generally, given a graph, we seek the 'simplest' orientable surface on which the graph can be embedded. With this in mind, we make the following definition.

> **Definition**
>
> The **orientable genus** $g(G)$ of a graph G is the smallest value of h for which there is a simple embedding of G on the orientable surface O_h.

For example, if G is a planar graph, then $g(G) = 0$, while for the non-planar graphs K_5 and $K_{3,3}$ we have $g(K_5) = g(K_{3,3}) = 1$.

Worked problem 3.1

By finding a suitable embedding on a torus, show that the graph $K_{3,4}$ has orientable genus 1.

Solution

Since $K_{3,4}$ is non-planar, the orientable genus is at least 1. Figure 3.5 shows that $K_{3,4}$ can be embedded on a torus. Thus, $g(K_{3,4}) = 1$. ∎

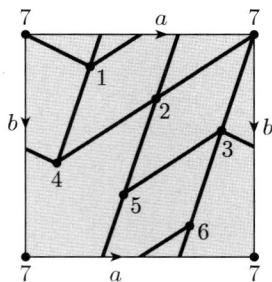

Figure 3.5 The '3-set' is $\{1, 2, 3\}$ and the '4-set' is $\{4, 5, 6, 7\}$.

Problem 3.2

By finding a suitable embedding on a torus, show that the graph K_6 has orientable genus 1.

Hint Begin with the first of the embeddings of K_5 illustrated in Figure 3.1.

The graph K_7 can also be embedded on a torus, as shown in Figure 3.6, and so $g(K_7) = 1$. However, the complete graph K_8 cannot be embedded on a torus, as we shall see, although it can be embedded on a 2-fold torus; thus, $g(K_8) = 2$. But what about other graphs? We return to this question shortly.

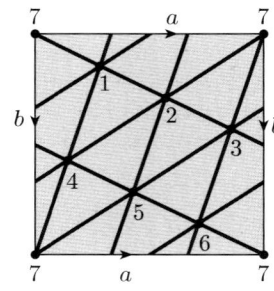

Figure 3.6

Embeddings on non-orientable surfaces

Just as we can embed graphs on orientable surfaces, such as the torus, so we can also embed them on non-orientable surfaces, such as the projective plane or the Klein bottle. Figure 3.7(a) shows how the complete graph K_6 can be embedded on a projective plane. Figure 3.7(b) shows how the complete bipartite graph $K_{4,4}$ can be embedded on a Klein bottle.

The '4-sets' are $\{1, 2, 3, 4\}$ and $\{5, 6, 7, 8\}$.

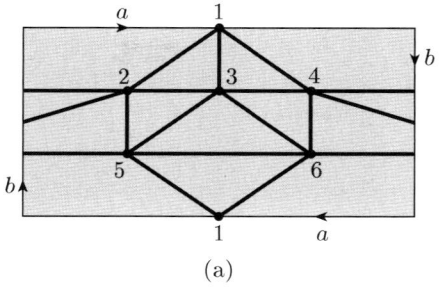
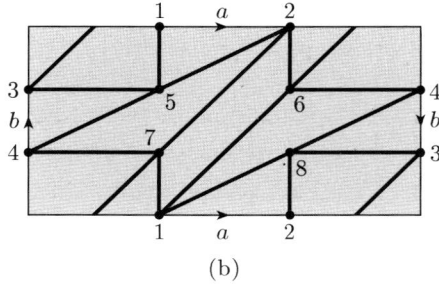

(a)   (b)

Figure 3.7

Problem 3.3

(a) Embed the complete graph K_5 on a projective plane.

(b) Embed the complete bipartite graph $K_{3,3}$ on a projective plane.

Recall from *Unit B3* that compact non-orientable surfaces without boundary can be classified as follows.

Unit B3, Theorem 4.5.

Theorem 3.2   Non-orientable surfaces

Every compact non-orientable surface without boundary is topologically equivalent to a sphere with k cross-caps, for some positive integer k.

Theorem 3.1 refers to 'some *non-negative* integer h', while Theorem 3.2 refers to 'some *positive* integer k': there is an *orientable* surface O_0 (the sphere), but no *non-orientable* surface N_0.

When $k = 1$ we have a projective plane; when $k = 2$ we have a Klein bottle. Any compact non-orientable surface that is topologically equivalent to a sphere with k cross-caps is called a **non-orientable surface of genus k**, denoted by N_k.

Just as in the orientable case, given a graph, we seek the 'simplest' non-orientable surface on which the graph can be embedded — the one with the fewest cross-caps. This prompts the following definition.

Definition

The **non-orientable genus** $q(G)$ of a graph G is the smallest value of k for which there is a simple embedding of G on the non-orientable surface N_k.

For example, $q(K_5) = q(K_6) = 1, q(K_{3,3}) = 1$ and, as we shall shortly see, $q(K_{4,4}) = 2$. But what about other graphs? We return to this question shortly.

3.2   The orientable genus

We now return to compact orientable surfaces without boundary. Using Euler's Formula for such surfaces, we obtain a lower bound for the orientable genus of certain graphs; surprisingly often this lower bound gives the actual value of the orientable genus, though this is very difficult to prove, as we shall see.

We saw earlier that a simple embedding of a connected graph on a surface without boundary defines a subdivision of that surface. Thus we can apply Euler's Formula to such simple embeddings. In *Unit B1*, you saw that the Euler characteristic of the h-fold torus O_h is $2 - 2h$. Thus, Euler's Formula for connected graphs embedded on an orientable surface takes the following form.

Unit B1, Theorem 3.3.
We could also use Unit B3, Theorem 3.3.

Theorem 3.3   Euler's Formula for the orientable surface O_h

If a connected graph with V vertices, E edges and F faces has a simple embedding on the orientable surface O_h, then

$$V - E + F = 2 - 2h.$$

Euler's Formula does not hold in general for non-simple embeddings.

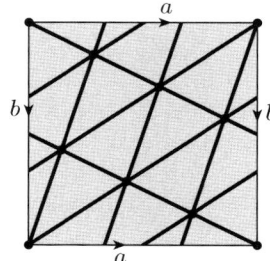

For example, the drawing of K_7 in Figure 3.8 is a simple embedding on the torus ($h = 1$) with $V = 7$, $E = 21$ and $F = 14$; in this case, Euler's Formula becomes

$$V - E + F = 7 - 21 + 14 = 0 \ (= 2 - 2h \text{ when } h = 1).$$

Problem 3.4

Verify Euler's Formula for the toroidal graphs in:

(a) Worked problem 3.1;   (b) Problem 3.2.

Figure 3.8

Using Euler's Formula, we can obtain lower bounds for the orientable genus of a given graph. We first need a preliminary result.

Lemma 3.4

Let G be a connected graph with no vertex of degree less than 3. In any simple embedding of G on a compact surface without boundary, let the subdivision so defined have F faces and E edges. Then:
(a) $F \leq \frac{2}{3}E$;
(b) if G has no triangles, $F \leq \frac{1}{2}E$.

Lemma 3.4(a) corresponds to Lemma 1.4(b) for maps.

Proof

(a) Each face of the corresponding subdivision has at least three edges, and no vertex has degree less than 3. Each edge belongs to two faces, so on counting edges, we obtain $2E \geq 3F$. Thus $F \leq \frac{2}{3}E$.

(b) The proof of (b) is similar. ∎

Remark

Since no vertex has degree less than 3, we cannot have situations like that in Figure 3.9(a), where the face A meets itself at the edge e. Since the embedding is simple, each face is homeomorphic to an open disc, and we avoid situations like that in Figure 3.9(b) where the face B meets itself at the edge f.

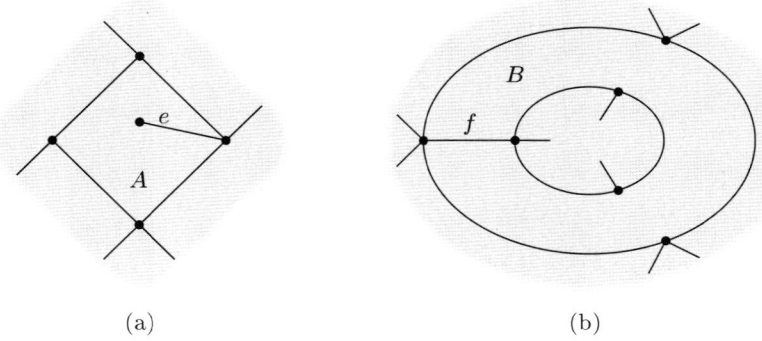

(a)   (b)

Figure 3.9

Problem 3.5

Prove Lemma 3.4(b).

We can now prove the promised result. Here, $\lceil x \rceil$ denotes the smallest integer greater than or equal to x; for example,

$$\lceil 2 \rceil = \lceil 1.234 \rceil = \lceil 1.678 \rceil = 2.$$

Thus, if x is not an integer, we round it up to the next integer.

> **Theorem 3.5   Lower bounds for the orientable genus of a graph**
>
> Let G be a connected graph with no vertex of degree less than 3, and with V vertices and E edges.
>
> (a) The orientable genus of G satisfies
>
> $$g(G) \geq \lceil \tfrac{1}{6}(E - 3V + 6) \rceil.$$
>
> (b) If G has no triangles, the orientable genus of G satisfies
>
> $$g(G) \geq \lceil \tfrac{1}{4}(E - 2V + 4) \rceil.$$

Proof

(a) Suppose that G has orientable genus g, and that a simple embedding of G on the surface O_g has F faces, where $V - E + F = 2 - 2g$.   We use Theorem 3.3 here.

By Lemma 3.4(a), $F \leq \tfrac{2}{3}E$, and so

$$2 - 2g = V - E + F \leq V - E + \tfrac{2}{3}E = V - \tfrac{1}{3}E,$$

which can be rearranged to give

$$g \geq \tfrac{1}{6}(E - 3V + 6).$$

Since g is an integer, this gives

$$g(G) \geq \lceil \tfrac{1}{6}(E - 3V + 6) \rceil.$$

(b) If G has no triangles, then by Lemma 3.4(b), $F \leq \tfrac{1}{2}E$. Thus,

$$2 - 2g = V - E + F \leq V - E + \tfrac{1}{2}E = V - \tfrac{1}{2}E,$$

which can be rearranged and rounded up, as above, to give

$$g(G) \geq \lceil \tfrac{1}{4}(E - 2V + 4) \rceil. \qquad \blacksquare$$

Complete graphs

We know that the complete graph K_7 can be embedded on a torus, so $g(K_7) = 1$. We can now verify that the complete graph K_8 cannot be embedded on a torus. Since $V = 8$ and $E = 28$, Theorem 3.5(a) tells us that

$$g(K_8) \geq \lceil \tfrac{1}{6}(E - 3V + 6) \rceil = \lceil \tfrac{1}{6}(28 - 24 + 6) \rceil = \lceil \tfrac{10}{6} \rceil = 2,$$

and so the orientable genus of K_8 is at least 2. However, K_8 can be embedded on a 2-fold torus, as shown in Figure 3.10, so $g(K_8) = 2$.

You saw in *Unit B1* (Figure 1.33) that the 2-fold torus can be represented as an octagon with edges identified as shown.

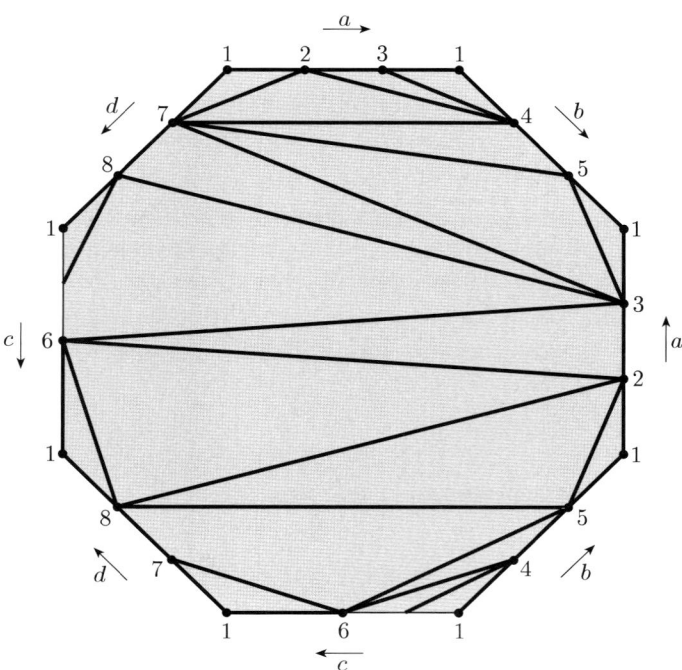

Figure 3.10

Using Theorem 3.5(a), we can similarly obtain a lower bound for the orientable genus of any complete graph. Since K_n has n vertices and $\tfrac{1}{2}n(n - 1)$ edges, we have:

$$g(K_n) \geq \lceil \tfrac{1}{6}(E - 3V + 6) \rceil = \lceil \tfrac{1}{6}(\tfrac{1}{2}n(n - 1) - 3n + 6) \rceil$$
$$= \lceil \tfrac{1}{12}(n^2 - 7n + 12) \rceil = \lceil \tfrac{1}{12}(n - 3)(n - 4) \rceil.$$

Surprisingly, this lower bound *always* gives the correct value for the orientable genus of a complete graph. This remarkable fact was proved by Gerhard Ringel and Ted Youngs in 1968.

Theorem 3.6 Ringel–Youngs Theorem for $g(K_n)$

The orientable genus of the complete graph K_n ($n \geq 3$) is

$$g(K_n) = \lceil \tfrac{1}{12}(n - 3)(n - 4) \rceil.$$

The requirement $n \geq 3$ arises from the need to use Theorem 3.5 (and thus Lemma 3.4) in the proof, which requires vertices of degree at least 3.

For example:

$$g(K_5) = \lceil \tfrac{1}{12}(5 - 3)(5 - 4) \rceil = \lceil \tfrac{2}{12} \rceil = 1;$$
$$g(K_6) = \lceil \tfrac{1}{12}(6 - 3)(6 - 4) \rceil = \lceil \tfrac{6}{12} \rceil = 1;$$
$$g(K_7) = \lceil \tfrac{1}{12}(7 - 3)(7 - 4) \rceil = \lceil \tfrac{12}{12} \rceil = 1;$$
$$g(K_8) = \lceil \tfrac{1}{12}(8 - 3)(8 - 4) \rceil = \lceil \tfrac{20}{12} \rceil = 2.$$

The proof of the Ringel–Youngs Theorem for $g(K_n)$, which we omit, is long and complicated and took many years to complete. It involves the consideration of twelve separate cases, depending on the remainder when n is divided by 12.

Problem 3.6

(a) Use the Ringel–Youngs Theorem to determine the orientable genus of the complete graph K_n, for $n = 9, 10, 11, 12$ and 13.

(b) What is the smallest positive integer that is not the orientable genus of any complete graph?

Complete bipartite graphs

The above methods can also be used to obtain a lower bound for the orientable genus of any complete bipartite graph $K_{r,s}$.

We first note that $K_{r,s}$ contains no triangles. To see this, suppose that v is any vertex in the r-set of vertices. Any triangle with v as a vertex must include two edges emerging from v. Each of these edges must end at a different vertex of the s-set of vertices. So to complete the triangle, we need an edge joining vertices in the s-set. Since there are no such edges, there is no triangle with v as a vertex. Similarly, there is no triangle involving any vertex w from the s-set. Hence $K_{r,s}$ contains no triangles.

Since $K_{r,s}$ has no triangles, we can use Theorem 3.5(b) to obtain a lower bound for the genus of any complete bipartite graph. As $K_{r,s}$ has rs edges and $r + s$ vertices, we have:

$$g(K_{r,s}) \geq \lceil \tfrac{1}{4}(E - 2V + 4) \rceil$$
$$= \lceil \tfrac{1}{4}(rs - 2(r+s) + 4) \rceil$$
$$= \lceil \tfrac{1}{4}(r-2)(s-2) \rceil.$$

For example, when $r = 3$, $s = 4$, we have

$$g(K_{3,4}) \geq \lceil \tfrac{1}{4}(3-2)(4-2) \rceil = \lceil \tfrac{1}{2} \rceil = 1;$$

in fact, as we saw in Worked problem 3.1, $g(K_{3,4}) = 1$.

Again, this lower bound *always* gives the correct value for the genus of a complete bipartite graph. The proof is very complicated, and so we omit it. It was completed in its entirety by Gerhard Ringel in 1965, three years before the proof of the Ringel–Youngs Theorem for $g(K_n)$.

Theorem 3.7 Ringel's Theorem for $g(K_{r,s})$

The orientable genus of the complete bipartite graph $K_{r,s}$ ($r, s \geq 3$) is

$$g(K_{r,s}) = \lceil \tfrac{1}{4}(r-2)(s-2) \rceil.$$

The requirement $r, s \geq 3$ arises from the need to use Theorem 3.5 (and thus Lemma 3.4) in the proof, which requires vertices of degree at least 3.

For example,

$$g(K_{4,4}) = \lceil \tfrac{1}{4}(4-2)(4-2) \rceil = \lceil 1 \rceil = 1.$$

Problem 3.7

Use Theorem 3.7 to determine the orientable genus of the complete bipartite graphs $K_{5,5}$, $K_{5,6}$, $K_{5,7}$ and $K_{5,8}$.

3.3 The non-orientable genus

We now turn our attention to compact non-orientable surfaces without boundary. Our treatment closely follows the above discussion for the orientable genus.

As in the orientable case, a simple embedding of a connected graph on a surface without boundary defines a subdivision of that surface, so we can apply Euler's Formula to such simple embeddings. Also, by Theorem 3.5 of *Unit B3*, the Euler characteristic of N_k, the sphere with k cross-caps, is $2 - k$. It follows that Euler's Formula for connected graphs embedded on a non-orientable surface takes the following form.

Theorem 3.8 Euler's Formula for the non-orientable surface N_k

If a connected graph with V vertices, E edges and F faces has a simple embedding on the non-orientable surface N_k, then

$$V - E + F = 2 - k.$$

Again, this formula does not hold in general for non-simple embeddings.

For example, the drawing of K_6 in Figure 3.11(a) is a simple embedding on the projective plane $(k = 1)$ with $V = 6$, $E = 15$, $F = 10$, and so

$$V - E + F = 6 - 15 + 10 = 1 \ (= 2 - k \text{ when } k = 1).$$

Similarly, the drawing of $K_{4,4}$ in Figure 3.11(b) is a simple embedding on the Klein bottle $(k = 2)$ with $V = 8$, $E = 16$, $F = 8$, and so

$$V - E + F = 8 - 16 + 8 = 0 \ (= 2 - k \text{ when } k = 2).$$

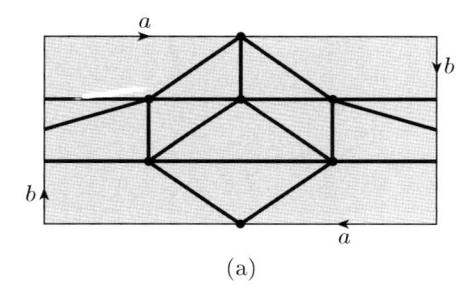

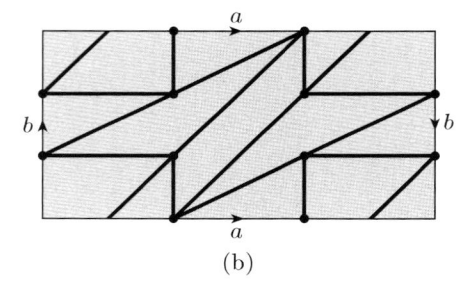

(a)   (b)

Figure 3.11

Problem 3.8

Verify Euler's Formula for each of the graphs in Problem 3.3.

We now obtain lower bounds for the non-orientable genus of a graph; the proof is similar to that of Theorem 3.5, and we leave the details to you as a problem.

> **Theorem 3.9 Lower bounds for the non-orientable genus of a graph**
>
> Let G be a connected graph with no vertex of degree less than 3, and with V vertices and E edges.
>
> (a) The non-orientable genus of G satisfies
> $$q(G) \geq \lceil \tfrac{1}{3}(E - 3V + 6) \rceil.$$
>
> (b) If G has no triangles, the non-orientable genus of G satisfies
> $$q(G) \geq \lceil \tfrac{1}{2}(E - 2V + 4) \rceil.$$

Problem 3.9

Prove Theorem 3.9.

We now show that the complete graph $K_{4,4}$ cannot be embedded on a projective plane. Theorem 3.9(b) tells us that

$$q(K_{4,4}) \geq \lceil \tfrac{1}{2}(E - 2V + 4) \rceil = \lceil \tfrac{1}{2}(16 - 16 + 4) \rceil = \lceil 2 \rceil = 2,$$

and so the non-orientable genus of $K_{4,4}$ is at least 2. However, $K_{4,4}$ can be embedded on a Klein bottle, as we saw in Subsection 3.1. Thus, $q(K_{4,4}) = 2$.

Recall that the complete bipartite graphs contain no triangles.

See Figure 3.7(b).

Complete graphs

Using Theorem 3.9(a), we can obtain a lower bound for the non-orientable genus of any complete graph. Since K_n has n vertices and $\tfrac{1}{2}n(n-1)$ edges, we have:

$$q(K_n) \geq \lceil \tfrac{1}{3}(E - 3V + 6) \rceil = \lceil \tfrac{1}{3}(\tfrac{1}{2}n(n-1) - 3n + 6) \rceil$$
$$= \lceil \tfrac{1}{6}(n^2 - 7n + 12) \rceil = \lceil \tfrac{1}{6}(n-3)(n-4) \rceil.$$

In fact, this lower bound gives the correct value for the non-orientable genus of a complete graph, *except for $n = 7$*, when the formula gives 2, but the correct value is 3. This remarkable result was proved by Gerhard Ringel in 1954. We omit the proof.

It is surprising that the non-orientable case was easier to settle than the orientable one.

> **Theorem 3.10 Ringel's Theorem for $q(K_n)$**
>
> The non-orientable genus of the complete graph K_n ($n \geq 3$) is
> $$q(K_n) = \lceil \tfrac{1}{6}(n-3)(n-4) \rceil,$$
> except that $q(K_7) = 3$.

The requirement $n \geq 3$ arises from the need to use Theorem 3.9 (and thus Lemma 3.4) in the proof, which requires vertices of degree at least 3.

For example,

$$q(K_5) = \lceil \tfrac{1}{6}(5-3)(5-4) \rceil = \lceil \tfrac{2}{6} \rceil = 1;$$
$$q(K_6) = \lceil \tfrac{1}{6}(6-3)(6-4) \rceil = \lceil \tfrac{6}{6} \rceil = 1.$$

Problem 3.10

Use Theorem 3.10 to determine the non-orientable genus of the complete graph K_n, for $n = 7, 8, 9$ and 10.

Complete bipartite graphs

Since $K_{r,s}$ has no triangles, we can use Theorem 3.9(b) to obtain a lower bound for the non-orientable genus of any complete bipartite graph. Since $K_{r,s}$ has rs edges and $r+s$ vertices, we have:

$$\begin{aligned} q(K_{r,s}) &\geq \lceil \tfrac{1}{2}(E - 2V + 4) \rceil \\ &= \lceil \tfrac{1}{2}(rs - 2(r+s) + 4) \rceil \\ &= \lceil \tfrac{1}{2}(r-2)(s-2) \rceil. \end{aligned}$$

As in the orientable case, this lower bound *always* gives the correct value for the non-orientable genus. This result is also due to Gerhard Ringel. Again we omit the proof.

Theorem 3.11   Ringel's Theorem for $q(K_{r,s})$

The non-orientable genus of the complete bipartite graph $K_{r,s}$ ($r, s \geq 3$) is

$$q(K_{r,s}) = \lceil \tfrac{1}{2}(r-2)(s-2) \rceil.$$

The requirement $r, s \geq 3$ arises from the need to use Theorem 3.9 (and thus Lemma 3.4) in the proof, which requires vertices of degree at least 3.

For example,

$$g(K_{3,3}) = \lceil \tfrac{1}{2}(3-2)(3-2) \rceil = \lceil \tfrac{1}{2} \rceil = 1.$$

Problem 3.11

Use Theorem 3.11 to determine the non-orientable genus of the complete bipartite graphs $K_{5,5}$, $K_{5,6}$, $K_{5,7}$ and $K_{5,8}$.

4 Colouring maps on surfaces

After working through this section, you should be able to:
- ▶ show that any map on a torus can be coloured with seven colours;
- ▶ show that any map on a projective plane can be coloured with six colours;
- ▶ define the *chromatic number* of a surface;
- ▶ state the *Ringel–Youngs Theorem* for the chromatic number of an orientable or non-orientable surface.

We now return to the colouring of maps. In Section 1 we stated the Four-colour Theorem for spherical maps, and we proved the Five-colour Theorem for such maps. We also stated that every map on a torus can be coloured with at most seven colours, and we gave an example of a toroidal map that requires seven colours. In this section we prove this result concerning the torus, and we extend our discussion to the colouring of maps on other compact surfaces. As you will see, our results are closely related to the results on the orientable and non-orientable genus of the complete graph, discussed in Section 3.

4.1 Maps on a torus and a projective plane

In 1890, in the same paper in which he presented the error in Kempe's proof of the Four-colour Theorem and salvaged enough to prove the Five-colour Theorem, Heawood attempted to generalize map-colouring to maps on surfaces other than a sphere. In particular, he showed that, just as *four* seemed to be the 'chromatic number' for maps on a sphere, so *seven* is the number for a torus, in the sense that:

- every map on a torus can be coloured with at most seven colours;
- there is at least one map on a torus that requires seven colours.

In this subsection we justify this result for maps on a torus, and then obtain the corresponding result for maps on a projective plane. Many of the results are exact analogues of the corresponding results for a sphere, and in these cases we ask you to fill in the details.

An example of a toroidal map that needs seven colours was given in Figure 1.8, and a variant is shown in Figure 4.1.

We begin by noting that a map on a compact surface is a subdivision, and hence is a connected graph with no crossings. Therefore a map on a compact surface is a simple embedding of a connected graph on that connected surface. This equivalence allows us to combine results for maps from Section 1 with results for graphs from Section 3.

Maps on a torus

To prove that every map on a torus can be coloured with seven colours, we start with Euler's Formula for a torus.

> **Theorem 4.1 Euler's Formula for a torus**
>
> If a map with V vertices, E edges and F faces has a simple embedding on a torus, then
>
> $$V - E + F = 0.$$

This is a special case of Theorem 3.3.

The next step is to show that every map on a torus must contain at least one triangle, quadrilateral, pentagon or hexagon.

> **Theorem 4.2**
>
> Every toroidal map has at least one face that meets six or fewer other faces.

This result is the analogue for a torus of Theorem 1.5 for a sphere.

Problem 4.1

Prove Theorem 4.2, by adapting the proof of Theorem 1.5.

We can now deduce that all maps on a torus can be coloured with at most seven colours.

> **Theorem 4.3 Seven-colour Theorem for toroidal maps**
>
> The faces of every toroidal map can be coloured with at most seven colours in such a way that any two faces with a common edge are assigned different colours.

Problem 4.2

Prove Theorem 4.3, by adapting the proof of Theorem 1.6.

We have already seen (Figure 1.8) that there is at least one toroidal map that requires seven colours, so Theorem 4.3 cannot be improved.

There is a useful connection between these results and those of Section 3. Figure 4.1 shows a toroidal map that requires seven colours. When we take the dual of this map, we obtain a seven-colouring of a simple embedding of K_7 on a torus. Reversing the argument shows that a seven-colouring of a simple embedding of K_7 on a torus leads to a toroidal map that requires seven colours. Furthermore, since we know from Section 2 that K_7 has chromatic number 7, this provides further demonstration that we cannot improve on Theorem 4.3.

It is the simple embedding of K_7 on a torus shown in Figures 3.6 and 3.8.

seven-coloured map   taking the dual   simple embedding of K_7

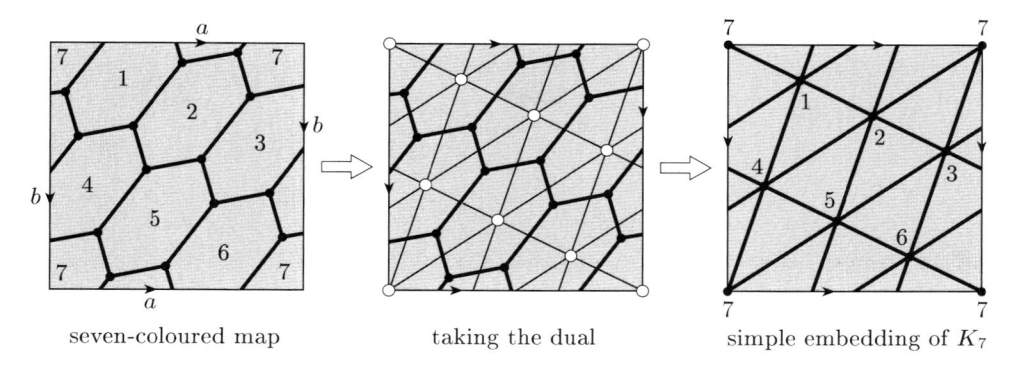

Figure 4.1

Maps on a projective plane

We can adapt the above discussion to maps on a projective plane.

Just as *seven* is the 'chromatic number' for toroidal maps, so *six* is the number for maps on a projective plane.

To prove that every map on a projective plane can be coloured with at most six colours, we again start with Euler's Formula.

> *Theorem 4.4   Euler's Formula for a projective plane*
>
> If a map with V vertices, E edges and F faces has a simple embedding on a projective plane, then
>
> $V - E + F = 1.$

This is a special case of Theorem 3.8.

The next step is to show that every map on a projective plane must contain at least one triangle, quadrilateral or pentagon.

> *Theorem 4.5*
>
> Every map on a projective plane has at least one face that meets five or fewer other faces.

We omit the proof, which is similar to those of Theorems 1.5 and 4.2.

We can now deduce that all maps on a projective plane can be coloured with at most six colours.

> *Theorem 4.6   Six-colour Theorem for maps on a projective plane*
>
> The faces of every map on a projective plane can be coloured with at most six colours in such a way that any two faces with a common edge are assigned different colours.

We omit the proof, which is similar to those of Theorems 1.6 and 4.3.

An example of a map that needs six colours is shown in Figure 4.2. Thus Theorem 4.6 cannot be improved.

There is again a useful connection between these results and those of Section 3. Figure 4.3 shows that, when we take the dual of the map in Figure 4.2, we obtain a six-colouring of a simple embedding of K_6 on the projective plane. Reversing the argument shows that a six-colouring of a simple embedding of K_6 on a projective plane leads to a map on a projective plane that requires six colours. Furthermore, since we know from Section 2 that K_6 has chromatic number 6, this confirms that we cannot improve on Theorem 4.6.

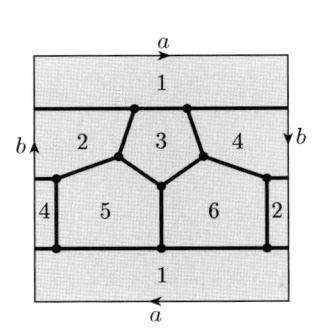

Figure 4.2

It is the same map as in Figure 4.2.

six-coloured map   taking the dual   simple embedding of K_6

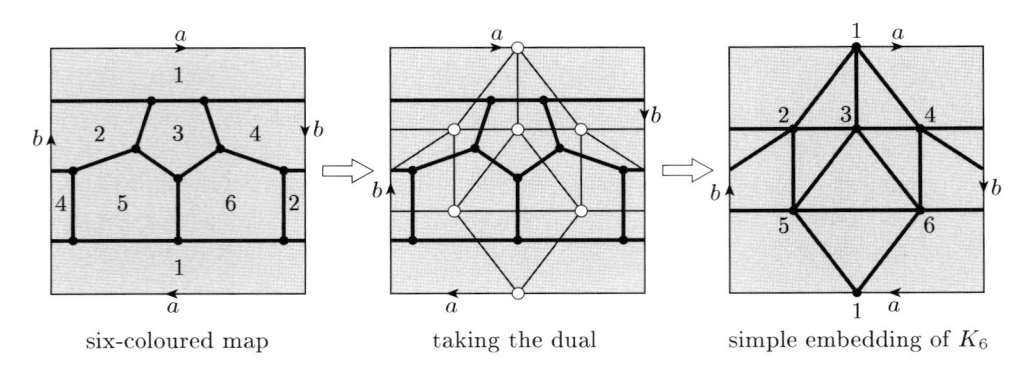

Figure 4.3

4.2 Maps on orientable surfaces

We now show how the examples in Subsection 4.1 can be generalized to maps on arbitrary orientable compact surfaces without boundary.

We begin by making the following definition, which applies to orientable and non-orientable surfaces.

> **Definition**
>
> The **chromatic number** $\mathrm{ch}(S)$ of a surface S is the smallest number of colours that are required to colour all maps on S.

Thus, the chromatic number of the sphere is 4 (by the Four-colour Theorem), and by the results of the previous subsection, the chromatic number of the torus is 7, and of the projective plane is 6.

You may like to compare this definition with the definition of the chromatic number of a graph. The duality of maps and graphs implies that the chromatic number $\mathrm{ch}(S)$ of a surface S is the greatest possible chromatic number $\mathrm{ch}(G)$ of a graph G that is the dual of a map on S.

See Subsection 2.3.

In his 1890 paper Heawood attempted to determine the chromatic number of any orientable surface O_h. He produced a formula, depending on h, which he claimed to be the chromatic number $\mathrm{ch}(O_h)$ of O_h, for $h \geq 1$.

> **Heawood's Formula**
>
> The chromatic number of the orientable surface O_h ($h \geq 1$) is
>
> $$\mathrm{ch}(O_h) = \left\lfloor \tfrac{1}{2}\left(7 + \sqrt{1 + 48h}\right) \right\rfloor.$$

$\lfloor x \rfloor$ is the 'integer part', or the 'rounded down' value, of x: for example, $\lfloor 8 \rfloor = \lfloor 8.123 \rfloor = \lfloor 8.999 \rfloor = 8$.

For example, the chromatic number of a torus ($h = 1$) is
$$\mathrm{ch}(O_1) = \lfloor \tfrac{1}{2}\left(7 + \sqrt{1+48}\right) \rfloor = \lfloor \tfrac{1}{2}(7+7) \rfloor = \lfloor 7 \rfloor = 7,$$
and the chromatic number of a 2-fold torus ($h = 2$) is
$$\mathrm{ch}(O_2) = \lfloor \tfrac{1}{2}\left(7 + \sqrt{1+96}\right) \rfloor = \lfloor \tfrac{1}{2}(7 + 9.8488...) \rfloor = \lfloor 8.4244... \rfloor = 8.$$

Using methods similar to those we used for the torus (where $h = 1$), Heawood proved correctly that every map on O_h (where $h \geq 1$) can be coloured with at most $\lfloor \tfrac{1}{2}\left(7 + \sqrt{1+48h}\right) \rfloor$ colours. Unfortunately, he was unable to prove (other than for the torus) that there are maps on O_h that require this number of colours. This general claim came to be known as the *Heawood Conjecture*: it took 78 years to prove, and was eventually completed by Ringel and Youngs as a consequence of their work on embedding graphs on orientable surfaces.

Problem 4.3

Use Heawood's Formula to calculate $\mathrm{ch}(O_h)$ for $h = 3, 4, 5$ and 6.

We now prove the result that Heawood proved:

every map on O_h ($h \geq 1$) can be coloured with at most $\lfloor \tfrac{1}{2}\left(7 + \sqrt{1+48h}\right) \rfloor$ colours.

We then show how the truth of the Heawood Conjecture follows from the Ringel–Youngs Theorem for $g(K_n)$ (Theorem 3.6).

Before proceeding, we need an algebraic result.

Lemma 4.7

If $x = \tfrac{1}{2}\left(7 + \sqrt{1+48h}\right)$, then $6 + \dfrac{12h - 12}{x} = x - 1.$

Proof If $x = \tfrac{1}{2}\left(7 + \sqrt{1+48h}\right),$

then $\quad x^2 = \tfrac{1}{4}\left(49 + (1 + 48h) + 14\sqrt{1+48h}\right) = \tfrac{25}{2} + 12h + \tfrac{7}{2}\sqrt{1+48h},$

so $\quad x^2 - 7x = \left(\tfrac{25}{2} + 12h + \tfrac{7}{2}\sqrt{1+48h}\right) - \tfrac{7}{2}\left(7 + \sqrt{1+48h}\right) = 12h - 12.$

Dividing by x and rearranging, we obtain
$$6 + \frac{12h - 12}{x} = x - 1. \qquad \blacksquare$$

We can now prove the following preliminary result.

Theorem 4.8

Every map on O_h ($h \geq 1$) has at least one face that meets fewer than $\lfloor \tfrac{1}{2}\left(7 + \sqrt{1+48h}\right) \rfloor$ other faces.

The method of proof is similar to those of Theorems 1.5 and 4.2.

Proof Suppose that a map M on the orientable surface O_h has V vertices, E edges and F faces.

Let $x = \frac{1}{2}\left(7 + \sqrt{1 + 48h}\right)$. We may assume that $F > x$, since the result is clearly true if there are x or fewer faces.

Substituting the inequality $V \leq \frac{2}{3}E$ from Lemma 1.4(a) into Euler's Formula for O_h (Theorem 3.3), we obtain

$$2 - 2h = V - E + F \leq \tfrac{2}{3}E - E + F,$$

which can be rearranged to give

$$E \leq 3F + 6h - 6.$$

Hence, on multiplying by $2/F$:

$$\frac{2E}{F} \leq \frac{6F + 12h - 12}{F} = 6 + \frac{12h - 12}{F}. \tag{4.1}$$

We now need to consider separately the cases $h = 1$ and $h > 1$.

Case 1 $h = 1$

Here, since $12h - 12 = 0$, inequality (4.1) reduces to $2E/F \leq 6$. Since $x = \lfloor x \rfloor = 7$, we conclude that

$$\frac{2E}{F} < \lfloor x \rfloor.$$

Case 2 $h > 1$

Here, $12h - 12 > 0$. Since we may assume $F > x$ (see above), it follows that $\dfrac{12h - 12}{F} < \dfrac{12h - 12}{x}$. Combining this with inequality (4.1) gives

$$\begin{aligned}\frac{2E}{F} &< 6 + \frac{12h - 12}{x} \\ &= x - 1 \quad \text{(by Lemma 4.7)} \\ &< \lfloor x \rfloor.\end{aligned}$$

Since in both cases $2E/F$ (the *average* number of edges per face) is less than $\lfloor x \rfloor$, it follows that *at least one* face must have fewer than $\lfloor x \rfloor = \left\lfloor \frac{1}{2}\left(7 + \sqrt{1 + 48h}\right) \right\rfloor$ edges. Thus, at least one face must meet fewer than $\left\lfloor \frac{1}{2}\left(7 + \sqrt{1 + 48h}\right) \right\rfloor$ other faces. ∎

Using Theorem 4.8, we now prove that all maps on O_h ($h \geq 1$) can be coloured with at most $\left\lfloor \frac{1}{2}\left(7 + \sqrt{1 + 48h}\right) \right\rfloor$ colours.

Theorem 4.9 Colouring Theorem for orientable surfaces

The faces of every map on the orientable surface O_h ($h \geq 1$) can be coloured with at most $\left\lfloor \frac{1}{2}\left(7 + \sqrt{1 + 48h}\right) \right\rfloor$ colours in such a way that any two faces with a common edge are assigned different colours.

The method of proof is similar to those of Theorems 1.6 and 4.3.

Proof The proof is by mathematical induction on the number of faces.

Let $x = \frac{1}{2}\left(7 + \sqrt{1 + 48h}\right)$. The result is clearly true for maps on O_h with up to $\lfloor x \rfloor$ faces.

We now assume that

all maps on O_h with k faces can be coloured with at most $\lfloor x \rfloor$ colours,   This is our induction hypothesis.

and show that

all maps on O_h with $k + 1$ faces can be so coloured.

Let M be a map on O_h with $k + 1$ faces. By Theorem 4.8, M has a face A that meets fewer than $\lfloor x \rfloor$ other faces. If we remove one of the edges e bounding A (say the edge separating faces A and B), then the resulting map N has k faces, one of which (let us call it C) replaces faces A and B of M (Figure 4.4). By our induction hypothesis, we can colour the faces of the map N with $\lfloor x \rfloor$ colours in such a way that neighbouring faces are coloured differently.

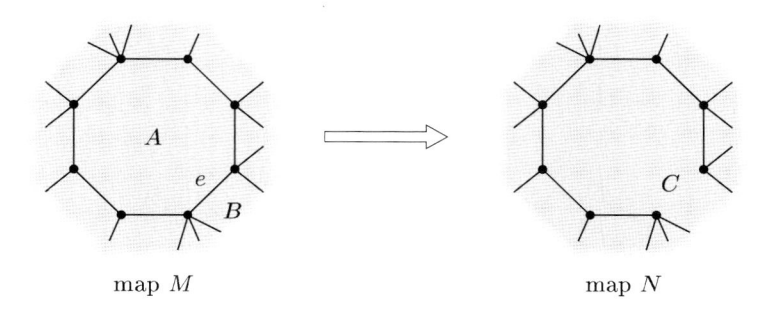

Figure 4.4

We now reinstate the edge e, and colour M as follows. For the faces other than A and B, we use the colours of the corresponding faces of N. For face B, we use the colour that was used on face C of N. Now the only face left to colour is A. Since A has fewer than $\lfloor x \rfloor$ neighbours, and $\lfloor x \rfloor$ colours are available, there is a spare colour that can be used. This gives a $\lfloor x \rfloor$-colouring of the faces of the original map M.

The result now follows by mathematical induction. ∎

We have thus proved the first result that Heawood proved:

every map on O_h ($h \geq 1$) can be coloured with at most $\left\lfloor \frac{1}{2}\left(7 + \sqrt{1 + 48h}\right) \right\rfloor$ colours.

The result that completes the proof of the Heawood Conjecture is that

there is a map on O_h ($h \geq 1$) that requires $\left\lfloor \frac{1}{2}\left(7 + \sqrt{1 + 48h}\right) \right\rfloor$ colours.

This was deduced by Ringel and Youngs from their formula for the orientable genus of the complete graph: Theorem 3.6.

$$g(K_n) = \left\lceil \tfrac{1}{12}(n-3)(n-4) \right\rceil.$$

To indicate how they did so, we recall from Section 2 that the chromatic number of the graph K_n is n: that is, n colours are needed to colour the vertices of the graph K_n so that adjacent vertices are coloured differently. It follows from the duality of maps and graphs that, if K_n can be embedded on the surface O_h, then its dual requires n colours: that is, $\text{ch}(O_h) \geq n$.

Let $n = \left\lfloor \tfrac{1}{2}\left(7 + \sqrt{1 + 48h}\right) \right\rfloor$.

Then, by the Ringel–Youngs Theorem for $g(K_n)$,

$$\begin{aligned}g(K_n) &= \lceil \tfrac{1}{12}(n-3)(n-4) \rceil \\ &= \lceil \tfrac{1}{12} \lfloor \tfrac{1}{2}(\sqrt{1+48h}+1) \rfloor \lfloor \tfrac{1}{2}(\sqrt{1+48h}-1) \rfloor \rceil \\ &\leq \lceil \tfrac{1}{48}(\sqrt{1+48h}+1)(\sqrt{1+48h}-1) \rceil \qquad \text{This line follows since } \lfloor x \rfloor \leq x \text{ for all } x. \\ &= \lceil \tfrac{1}{48}(1+48h-1) \rceil = h.\end{aligned}$$

It follows that K_n can be embedded on O_h, and hence that

$$\mathrm{ch}(O_h) \geq n = \left\lfloor \tfrac{1}{2}\left(7+\sqrt{1+48h}\right) \right\rfloor.$$

Combining this with Theorem 4.9, which tells us that

$$\mathrm{ch}(O_h) \leq n = \left\lfloor \tfrac{1}{2}\left(7+\sqrt{1+48h}\right) \right\rfloor,$$

we deduce the Ringel–Youngs Theorem for orientable surfaces.

Theorem 4.10 Ringel–Youngs Theorem for orientable surfaces

The chromatic number of the orientable surface O_h ($h \geq 1$) is

$$\mathrm{ch}(O_h) = \left\lfloor \tfrac{1}{2}\left(7+\sqrt{1+48h}\right) \right\rfloor.$$

Remark

The above *proof* of this theorem is valid only when $h \geq 1$. This is because we need $h \geq 1$ to prove Theorem 4.8. However, by the Four-colour Theorem, the *statement* is also true when $h = 0$, since the formula in this case reduces to $\left\lfloor \tfrac{1}{2}(7+\sqrt{1}) \right\rfloor = \lfloor 4 \rfloor = 4$.

If $h = 0$, then the condition $2E/F < \lfloor x \rfloor$ that is required in the proof of Theorem 4.8 does not follow.

We conclude this subsection with a 'practical application' of this result.

Mrs Ringel and the traffic cop

Gerhard Ringel visited California for the academic year 1967–68 to work with Ted Youngs on the proof of the Heawood Conjecture. Shortly after they were successful and the proof had been announced, Mrs Ringel was driving along the California expressway, and was stopped by a traffic cop for a traffic violation. On hearing that the culprit's name was Ringel, the traffic cop asked: 'Was your husband the one that solved the Heawood Conjecture?' Mrs Ringel, surprised, admitted that he was, and she was let off with only a warning. It transpired that the traffic cop's son had been in Ted Youngs' calculus class when the proof of the Heawood Conjecture had been announced.

4.3 Maps on non-orientable surfaces

We now adapt these ideas to the colouring of maps on any non-orientable compact surface without boundary.

The result we aim to prove is the non-orientable analogue of Heawood's Formula.

> The chromatic number of the non-orientable surface N_k ($k \geq 1$) is
> $$\text{ch}(N_k) = \left\lfloor \tfrac{1}{2} \left(7 + \sqrt{1+24k}\right) \right\rfloor, \quad (4.2)$$
> except for the Klein bottle ($k = 2$), for which $\text{ch}(N_2) = 6$.

Recall that there is no non-orientable surface with $k = 0$, so we do not have the $k = 0$ case to prove separately, as in the orientable case. Thus, once we have proved this result, we have dealt with all non-orientable compact surfaces without boundary.

For example, the chromatic number of a projective plane ($k = 1$) is
$$\text{ch}(N_1) = \left\lfloor \tfrac{1}{2}\left(7+\sqrt{1+24k}\right)\right\rfloor = \left\lfloor \tfrac{1}{2}\left(7+\sqrt{25}\right)\right\rfloor = \lfloor \tfrac{1}{2}(7+5)\rfloor = \lfloor 6 \rfloor = 6.$$

Problem 4.4

Use formula (4.2) to calculate $\text{ch}(N_k)$ for $k = 3, 4, 5$ and 6.

Using methods similar to those we used for orientable surfaces, we can prove that:

> every map on N_k ($k \geq 1$) can be coloured with at most $\left\lfloor \tfrac{1}{2}\left(7+\sqrt{1+24k}\right)\right\rfloor$ colours.

We then show how formula (4.2) follows from Ringel's Theorem for $q(K_n)$ (Theorem 3.10).

Before proceeding, we need the following analogue of Lemma 4.7.

> **Lemma 4.11**
>
> If $y = \tfrac{1}{2}\left(7+\sqrt{1+24k}\right)$, then $6 + \dfrac{6k-12}{y} = y - 1$.

The proof is almost identical to that of Lemma 4.7, so we omit it.

We can now prove the following preliminary result.

> **Theorem 4.12**
>
> Every map on N_k ($k \geq 1$) has at least one face that meets fewer than $\left\lfloor \tfrac{1}{2}\left(7+\sqrt{1+24k}\right)\right\rfloor$ other faces.

Proof If $k=1$, then $\lfloor \frac{1}{2}(7 + \sqrt{1+24k})\rfloor = \lfloor 6 \rfloor = 6$, and the result follows from Theorem 4.5.

Suppose next that $k \geq 2$ and we are given a map M on the non-orientable surface N_k. We may suppose that the map has V vertices, E edges and F faces.

Let $y = \frac{1}{2}(7 + \sqrt{1+24k})$. We may assume that $F > y$, since the result is clearly true if there are y or fewer faces.

Substituting the inequality $V \leq \frac{2}{3}E$ from Lemma 1.4(a) into Euler's Formula for N_k (Theorem 3.8), we obtain

$$2 - k = V - E + F \leq \tfrac{2}{3}E - E + F,$$

which can be rearranged to give

$$E \leq 3F + 3k - 6.$$

Hence, on multiplying by $2/F$,

$$\frac{2E}{F} \leq \frac{6F + 6k - 12}{F} = 6 + \frac{6k - 12}{F}. \tag{4.3}$$

We now need to consider separately the cases $k = 2$ and $k > 2$.

Case 1 $k = 2$

Here, since $6k - 12 = 0$, inequality (4.3) reduces to $2E/F \leq 6$. Since $y = \lfloor y \rfloor = 7$, we conclude that

$$\frac{2E}{F} < \lfloor y \rfloor.$$

Notice that, in proving Case 1 ($k = 2$), we have also shown that every map on N_2 has at least one face that meets fewer than 6 other faces.

Case 2 $k > 2$

Here, $6k - 12 > 0$. Since we may assume $F > y$ (see above), it follows that $(6k - 12)/F < (6k - 12)/y$. Combining this with inequality (4.3) gives

$$\frac{2E}{F} < 6 + \frac{6k - 12}{y}$$
$$= y - 1 \quad \text{(by Lemma 4.11)}$$
$$< \lfloor y \rfloor.$$

Since in both cases $2E/F$ (the *average* number of edges per face) is less than $\lfloor y \rfloor$, it follows that *at least one* face must have fewer than $\lfloor y \rfloor = \lfloor \frac{1}{2}(7 + \sqrt{1+24k})\rfloor$ edges. Thus, at least one face must meet fewer than $\lfloor \frac{1}{2}(7 + \sqrt{1+24k})\rfloor$ other faces. ∎

Using Theorems 4.6 and 4.12, we can prove that all maps on N_k ($k \geq 1$) can be coloured with at most $\lfloor \frac{1}{2}(7 + \sqrt{1+24k})\rfloor$ colours.

Theorem 4.13 *Colouring Theorem for non-orientable surfaces*

The faces of every map on the non-orientable surface N_k ($k \geq 1$) can be coloured with at most $\lfloor \frac{1}{2}(7 + \sqrt{1+24k})\rfloor$ colours in such a way that any two faces with a common edge are assigned different colours.

We omit the proof, since the method of proof for $k \geq 2$ is almost identical to that of Theorem 4.9, while the case $k = 1$ is just Theorem 4.6.

In the light of Case 1 of Theorem 4.12, we can also prove that the faces of every map on the non-orientable surface N_2 can be coloured with at most six colours in such a way that any two faces with a common edge are assigned different colours.

Theorem 4.13 tells us that:

> *every map on N_k ($k \geq 1$) can be coloured with at most $\left\lfloor \frac{1}{2} \left(7 + \sqrt{1 + 24k}\right) \right\rfloor$ colours.*

We now deduce that (except when $k = 2$):

> *there is a map on N_k ($k \geq 1$) that requires $\left\lfloor \frac{1}{2} \left(7 + \sqrt{1 + 24k}\right) \right\rfloor$ colours,*

by using Ringel's Theorem for the non-orientable genus of the complete graph:

$$q(K_n) = \left\lceil \tfrac{1}{6}(n-3)(n-4) \right\rceil, \quad \text{when } n \neq 7.$$

Theorem 3.10.

To do so, we recall from Section 2 that n colours are needed to colour the vertices of the graph K_n so that adjacent vertices are coloured differently. It follows from the duality of maps and graphs that, if K_n can be embedded on the surface N_k, then its dual requires n colours: that is, $\mathrm{ch}(N_k) \geq n$.

Let $n = \left\lfloor \frac{1}{2} \left(7 + \sqrt{1 + 24k}\right) \right\rfloor$.

Now, when $k = 1$, $n = 6$, when $k = 2$ or 3, $n = 7$, and when $k > 3$, $n \geq 8$. So, when $k \neq 2$ and $k \neq 3$, by Ringel's Theorem for $q(K_n)$,

$$\begin{aligned} q(K_n) &= \left\lceil \tfrac{1}{6}(n-3)(n-4) \right\rceil \\ &= \left\lceil \tfrac{1}{6} \left\lfloor \tfrac{1}{2} \left(\sqrt{1+24k} + 1\right) \right\rfloor \left\lfloor \tfrac{1}{2} \left(\sqrt{1+24k} - 1\right) \right\rfloor \right\rceil \\ &\leq \left\lceil \tfrac{1}{24} \left(\sqrt{1+24k} + 1\right)\left(\sqrt{1+24k} - 1\right) \right\rceil \\ &= \left\lceil \tfrac{1}{24}\left((1+24k) - 1\right)\right\rceil = k. \end{aligned}$$

This line follows since $\lfloor x \rfloor \leq x$ for all x.

It follows that, for $k \neq 2$ and $k \neq 3$, K_n can be embedded on N_k, and hence that

$$\mathrm{ch}(N_k) \geq n = \left\lfloor \tfrac{1}{2} \left(7 + \sqrt{1 + 24k}\right) \right\rfloor.$$

Combining this with Theorem 4.13, which tells us that

$$\mathrm{ch}(N_k) \leq n = \left\lfloor \tfrac{1}{2} \left(7 + \sqrt{1 + 24k}\right) \right\rfloor,$$

we have therefore deduced formula (4.2) for $k \neq 2$ and $k \neq 3$. It remains to deal with the cases where $k = 2$ and $k = 3$.

When $k = 2$, we make use of the remark following Theorem 4.13. We first note that, by Ringel's Theorem for $q(K_n)$, $q(K_6) = \left\lceil \tfrac{1}{6}(6-3)(6-4) \right\rceil = 1$, so that K_6 can be embedded on N_k for all $k \geq 1$, and in particular on N_2. Thus $\mathrm{ch}(N_2) \geq 6$. The remark following Theorem 4.13 tells us that $\mathrm{ch}(N_2) \leq 6$. Hence we have proved the exceptional case where formula (4.2) does not hold: $\mathrm{ch}(N_2) = 6$.

The case where $k = 3$ makes use of the exceptional case from Ringel's Theorem for $q(K_n)$.

Problem 4.5

Prove that, when $k = 3$, $\text{ch}(N_k) \geq \left\lfloor \frac{1}{2}(7 + \sqrt{1 + 24k}) \right\rfloor$. Deduce that formula (4.2) holds for $k = 3$.

We have therefore shown that formula (4.2) holds for all $k \geq 1$ except $k = 2$, in which case $\text{ch}(N_2) = 6$. This is the Ringel–Youngs Theorem for non-orientable surfaces.

> **Theorem 4.14 Ringel–Youngs Theorem for non-orientable surfaces**
>
> The chromatic number of the non-orientable surface N_k ($k \geq 1, k \neq 2$) is
>
> $$\text{ch}(N_k) = \left\lfloor \frac{1}{2}\left(7 + \sqrt{1 + 24k}\right) \right\rfloor,$$
>
> and $\text{ch}(N_2) = 6$.

5 Proving the Four-colour Theorem

After working through this section, you should be able to:
▶ outline a proof of the Four-colour Theorem, explaining the terms *unavoidable set* and *reducible configuration*.

This section is not assessed.

In Section 1 we stated the Four-colour Theorem.

> **Four-colour Theorem for spherical maps**
> The faces of every spherical map can be coloured with at most four colours in such a way that any two faces with a common edge are assigned different colours.

Using the fact that every spherical map contains a face that meets five or fewer other faces, we proved a weaker result, that the faces of every spherical map can be coloured with at most *five* colours. Alfred Kempe tried to prove the Four-colour Theorem using ideas similar to those in our proof of the Five-colour Theorem. His 'proof' of 1879 proceeded as follows.

Alfred Kempe was a former Cambridge student of Arthur Cayley. He became a distinguished barrister and Treasurer of the Royal Society.

Kempe's 'proof'

The 'proof' is by mathematical induction on the number of faces in the map.

The result is clearly true for spherical maps with up to four faces.

We now assume that

all spherical maps with k faces can be coloured with four colours,

and show that

all spherical maps with $k + 1$ faces can be so coloured.

This is our induction hypothesis.

Let M be a spherical map with $k+1$ faces. By Theorem 1.5, M has a face A that meets at most five other faces. We now consider three cases.

Case 1 A is a triangle

Since A has three neighbours, and four colours are available, there is a spare colour that can be used to colour A (Figure 5.1). This gives a four-colouring of the faces of the original map M.

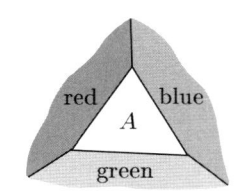

Figure 5.1

Case 2 A is a quadrilateral

We remove an edge e of A, colour the resulting k-faced map N, then reinstate e and colour all the faces of M except A (Figure 5.2). Since A has four neighbours and only four colours are available, there is no spare colour for A, whenever A meets faces of four different colours, as shown.

This procedure was spelled out in Theorems 1.6 and 1.7.

To overcome this difficulty, we consider the red and green neighbours of A, and investigate whether there is a path of red and green faces joining them.

Two situations can arise, as Figure 5.3 illustrates.

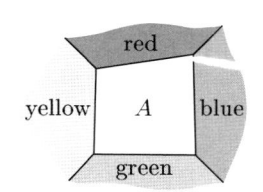

Figure 5.2

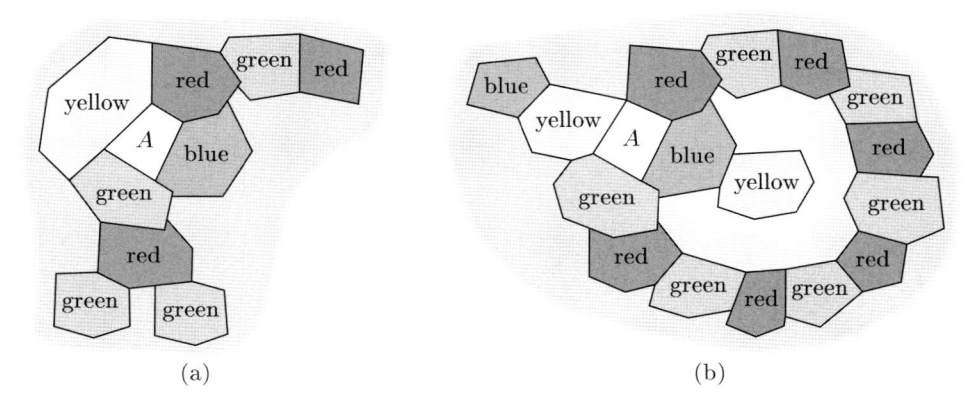

Figure 5.3

In case (a), all the red and green faces reachable from the red neighbour of A are different from those reachable from the green one, so there is no such red–green path of faces. In this case, we interchange the colours in the red–green part at the top, as shown in Figure 5.4(a). This replaces the red neighbour of A by a green one, so that A can now be coloured red. This completes the four-colouring of the original map M in this case.

In case (b), the red and green neighbours of A are linked by a red–green path of faces. In this case, there can be no blue–yellow path of faces between the blue and yellow neighbours of A, because the red–green path 'gets in the way'. We can therefore interchange the colours in the blue–yellow part on the right-hand side as shown in Figure 5.4(b). This replaces the blue neighbour of A by a yellow one, so that A can now be coloured blue. This completes the four-colouring of the original map M in this case.

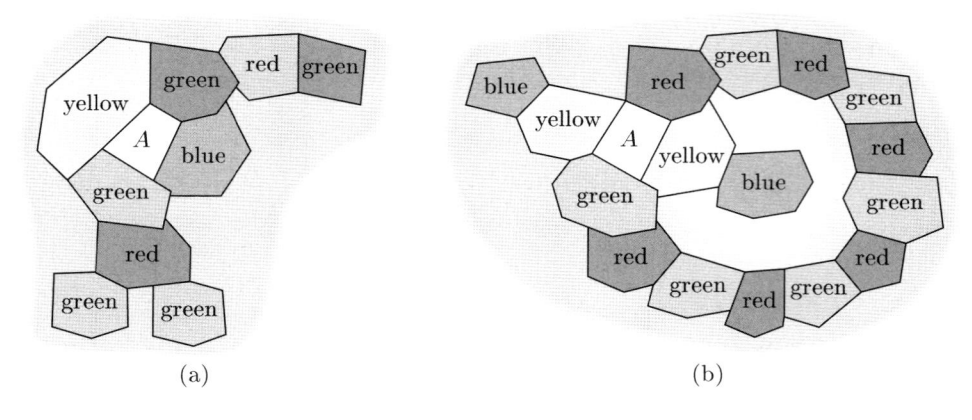

Figure 5.4

Case 3   A is a pentagon

As for Case 2, we remove an edge e of A, colour the resulting k-faced map N, then reinstate e and colour all the faces of M except A (Figure 5.5). Since A has five neighbours and only four colours are available, there is no spare colour for colouring A, whenever A meets faces of four different colours, as shown.

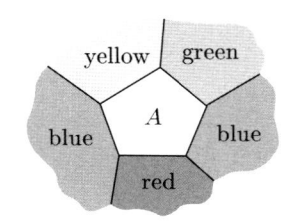

Figure 5.5

If there is no red–green path of faces joining the red and green neighbours of A or if there is no red–yellow path of faces joining the red and yellow neighbours of A, we proceed as in case (a) for the quadrilateral.

The difficulty arises if there is both a red–green path joining the red and green neighbours of A and a red–yellow path joining the red and yellow neighbours of A, as shown in Figure 5.6.

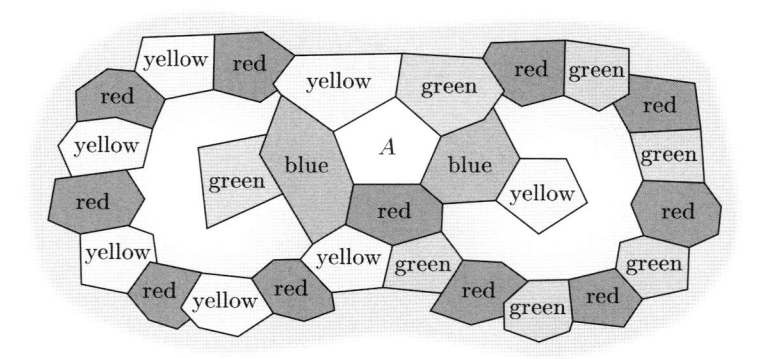

Figure 5.6

To overcome this difficulty Kempe carried out two colour interchanges around A at the same time. He interchanged the blues and yellows in the 'island' on the right, surrounded by A and the red–green path. He also interchanged the blues and greens in the 'island' on the left, surrounded by A and the red–yellow path. Now the only colours next to A are red, yellow and green. The face A can then be coloured blue. This gives a four-colouring of the original map M.

The result now follows by mathematical induction. ∎

The error in Kempe's 'proof'

In 1890, Heawood found a serious error in Kempe's proof, and constructed an example that demonstrates where Kempe's approach goes wrong.

Recall that Kempe tried to deal with *Case 3* above by carrying out two interchanges of colour at the same time. Either interchange is fine on its own, but the coloured map of Heawood, shown in Figure 5.7, demonstrates that it is sometimes impossible to perform both interchanges simultaneously to obtain the desired colouring.

Recall that our aim is to change the colours of the neighbours of the face A in such a way that there is a spare colour available for A.

Now, the blue and yellow neighbours of A are connected by a blue–yellow path that separates the reds and greens above A from those below A (Figure 5.8(a)). So we can interchange the reds and greens in the upper part without affecting those in the lower part, as shown in Figure 5.8(b).

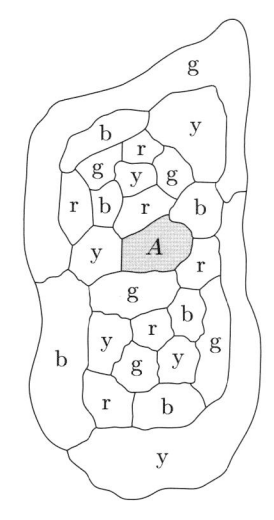

Figure 5.7

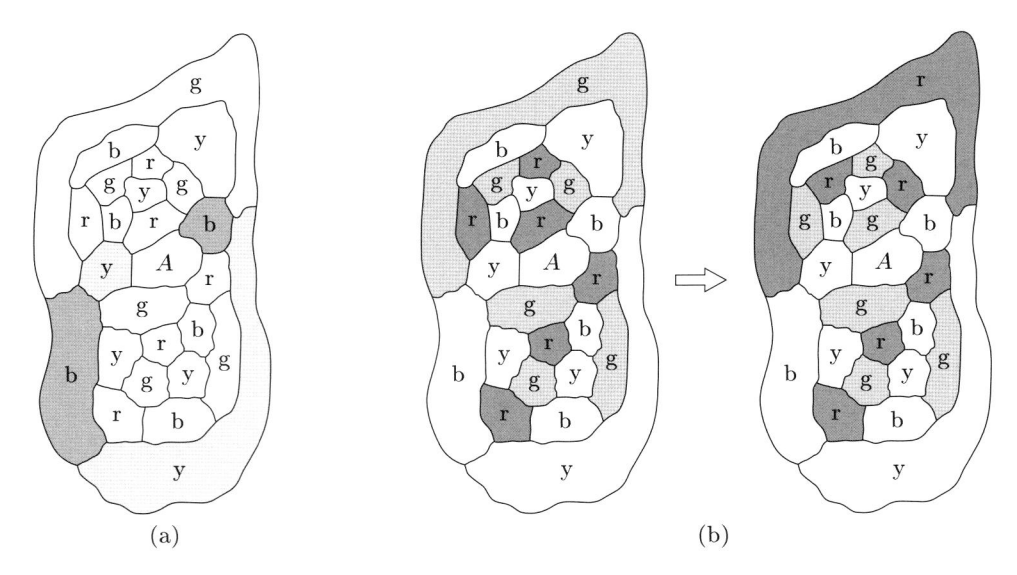

Figure 5.8

Alternatively, we can do another interchange of colour. The blue and green neighbours of A are connected by a blue–green path that separates the reds and yellows above A from those below A (Figure 5.9(a)). So we can interchange the reds and yellows in the lower part without affecting those in the upper part, as shown in Figure 5.9(b).

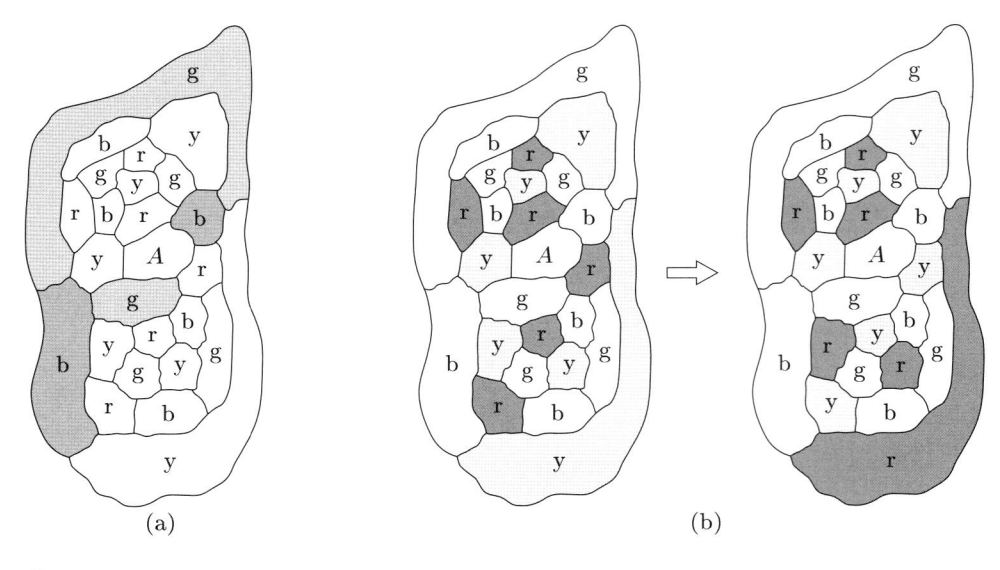

Figure 5.9

Either interchange is permissible on its own, but if we do both at the same time, we get two red countries coming together, as shown in Figure 5.10. So, in general, we cannot carry out two interchanges of colour at the same time.

The error in Kempe's argument was a major one, and it was many years before the Four-colour Theorem was proved.

Outline of a correct proof

The details of the eventual proof of 1976 are extremely complicated, but the basic ideas involved are simple and have their origins in Kempe's paper. These fundamental ideas, of an *unavoidable set* and a *reducible configuration*, were developed by a number of people during the ensuing years and led eventually to a solution of the problem. In what follows, we

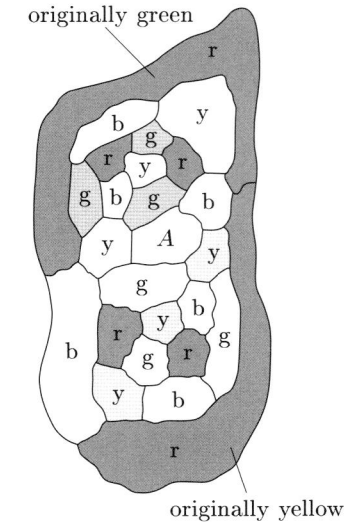

Figure 5.10

may assume (from Subsection 1.2) that our maps are cubic — that is, exactly three faces meet at each vertex.

Unavoidable sets

In Subsection 1.3 (Theorem 1.5) we proved that every spherical map must contain a triangle, a quadrilateral or a pentagon (Figure 5.11). We can express this by saying that these three types of face form an **unavoidable set** — this means if you look at any spherical map you cannot avoid finding at least one of them. Another unavoidable set for a spherical map consists of a triangle, a quadrilateral, *two* adjacent pentagons, and a pentagon adjacent to a hexagon (Figure 5.12). To prove this, we use the Face-counting Theorem (Theorem 1.3).

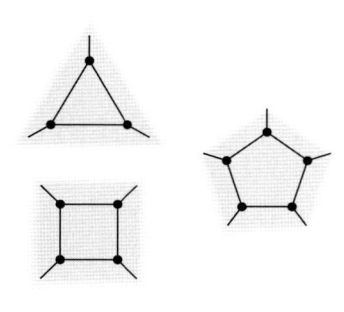

Figure 5.11

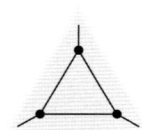

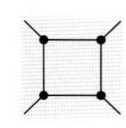

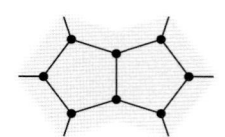

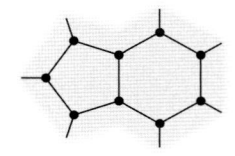

Figure 5.12

Theorem 5.1

The configurations in Figure 5.12 form an unavoidable set.

Proof The proof is by contradiction.

Suppose that there exists a spherical map M that contains none of the configurations in Figure 5.12. Thus, M has no triangles or quadrilaterals, and no pentagon can adjoin another pentagon or a hexagon. It follows that each pentagon can adjoin only faces with at least seven edges.

We now assign to each face an 'electrical charge':

to each face with k edges, we assign the charge $6 - k$.

Therefore:

- each pentagon ($k = 5$) is assigned a charge of 1;
- each hexagon ($k = 6$) is assigned a charge of 0;
- each heptagon ($k = 7$) is assigned a charge of -1;
- each octagon ($k = 8$) is assigned a charge of -2;
- each nonagon ($k = 9$) is assigned a charge of -3;

and so on, as Figure 5.13 illustrates.

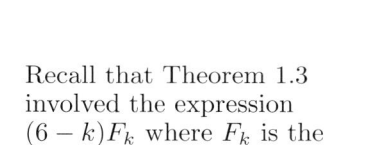

Recall that Theorem 1.3 involved the expression $(6 - k)F_k$ where F_k is the number of faces with k edges.

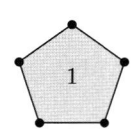

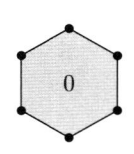

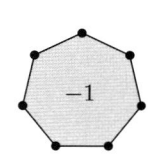

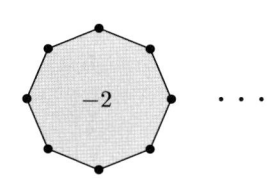

Figure 5.13

For each k, let F_k be the number of faces with k edges; then $F_3 = F_4 = 0$.

By Theorem 1.3, the total charge received by all the faces in M is then

$$3F_3 + 2F_4 + 1F_5 - 1F_7 - 2F_8 - 3F_9 - \cdots$$
$$= F_5 - F_7 - 2F_8 - 3F_9 - \cdots = 12.$$

Now transfer one-fifth of a unit of charge from each pentagon to its five negatively charged neighbours. Then the total charge on M remains 12, but:

- each pentagon now has zero charge;
- each hexagon still has zero charge;
- each heptagon, octagon, nonagon, ... still has negative charge.

This last statement follows since a heptagon can acquire non-negative charge only if it has at least five neighbouring pentagons, in which case at least two of these pentagons would have to be adjacent, which is not permissible. As for an octagon, it could acquire non-negative charge only if it were surrounded by at least ten neighbouring pentagons, which is a contradiction. The argument for nonagons, decagons, ... is similar.

Thus, each face now has negative or zero charge, and so the total charge on M cannot be 12. This contradiction establishes the result. ∎

Unavoidable sets have been found containing many hundreds of configurations of faces.

Reducible configurations

In his attempt to prove the Four-colour Theorem, Kempe showed that every triangle has the property that any four-colouring of the rest of the map can be extended to the triangle, since there is always a spare colour that can be used for the triangle. He also showed that the same can be said for a quadrilateral, although we may need to recolour some countries before we can recolour the quadrilateral. However, as we have seen, Kempe's argument does not work in the case of a pentagon.

A **reducible configuration** of faces in a map is an arrangement of faces with the property that every four-colouring of the rest of the map can be extended (either directly or after some recolouring) to a four-colouring of the countries in the configuration, and thereby to a four-colouring of the entire map. Thus, a triangle and a quadrilateral are examples of reducible configurations.

An example of a reducible configuration containing several faces is the *Birkhoff diamond*, shown in Figure 5.14.

> **Theorem 5.2**
>
> The Birkhoff diamond is a reducible configuration.

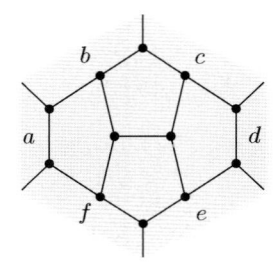

Figure 5.14

The Birkhoff diamond is named after the American mathematician George Birkhoff (1884–1944).

Outline of proof

If we remove the Birkhoff diamond from a map, we obtain a new map with fewer faces. By induction, we may assume that this new map is four-colourable, but can all such colourings be extended to the diamond?

To answer this, we list all the essentially different four-colourings of the six surrounding faces a, b, c, d, e, f:

abcdef	121234	121323*	123123	123142*	123214*	123243	123423
121212	121312	121324	123124	123143	123232	123412*	123424
121213*	121313*	121342*	123132*	123212*	123234*	123413	123432*
121232*	121314	121343	123134*	123213*	123242	123414	123434*

Sixteen of these colour schemes (indicated by *) can immediately be extended to the diamond: for example, the colour scheme 121213 can be extended as shown in Figure 5.15.

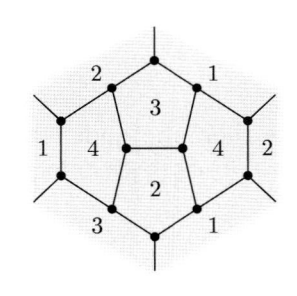

Figure 5.15

It can be verified that each of the remaining colour schemes can be extended to the diamond, by making one or more interchanges of colour. Thus all possible colour schemes can be extended to the Birkhoff diamond, and so it is a reducible configuration.

Combining the two ideas

The importance of the concepts of unavoidable sets and reducible configurations arises from the following observation.

> In order to prove the Four-colour Theorem, it is sufficient to find an unavoidable set of reducible configurations.

Since such a set is unavoidable, all spherical maps must contain at least one of the configurations; but, since all the configurations are reducible, any colouring of the rest of the map can be extended to any one of them, and thus the appropriate induction argument can be completed in each case. Another way of expressing this is to say that none of the configurations can appear in a counter-example to the Four-colour Theorem. Thus, there can be no counter-example, and so the theorem is true.

By the early 1970s large numbers of reducible configurations were known, as were methods for constructing unavoidable sets, but no one had been able to combine the ideas and find an unavoidable set of reducible configurations. Probabilistic arguments had indicated that such sets exist, but that they might contain thousands of configurations, far too many for the largest computers then available.

Eventually, in 1976, Kenneth Appel and Wolfgang Haken succeeded in constructing an unavoidable set of 1482 reducible configurations, thereby solving the 124-year-old problem. Their methods were largely those of the proofs of Theorems 5.1 and 5.2, although the technical details were far more complicated and involved massive reliance on a computer.

In particular, in developing methods for moving units of charge around the map (as in the proof of Theorem 5.1), they used much trial and error, together with a lot of insight and experience, and eventually obtained a systematic procedure that could be applied by hand. The configurations were then tested by computer for reducibility (as in our discussion of Theorem 5.2), and those that were not easily shown to be reducible were discarded and replaced by other configurations until an unavoidable set of reducible configurations was obtained, and the Four-colour Theorem was proved.

The achievement of Appel and Haken in settling the Four-colour Problem was a substantial one. At first some mathematicians were unhappy with the idea of a computer-aided proof but, as the underlying methods became clear, the solution gradually came to be accepted by the mathematical community.

Since then, methods for constructing unavoidable sets and testing for reducible configurations have improved greatly, the computer implementation has become more efficient, and computers have become ever faster — indeed, on modern high-speed machines, a proof now takes only a few hours to run.

Solutions to problems

1.1 **(a)** If we do not colour the exterior region, each map requires just three colours.

(i)

(ii)

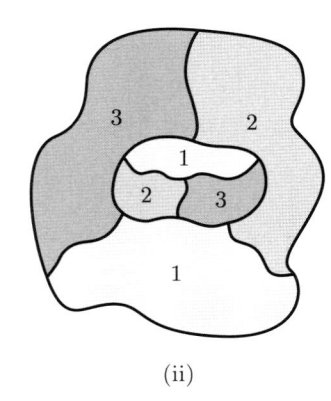

(b) If we do colour the exterior region, map (i) again requires just three colours, but map (ii) requires four colours.

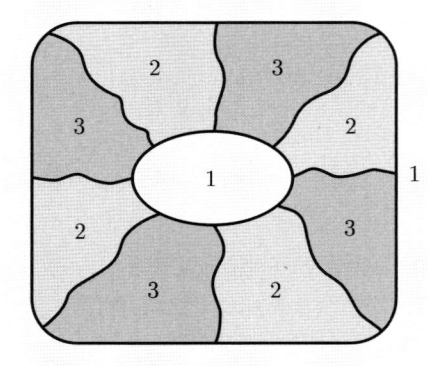

(i)

(ii)

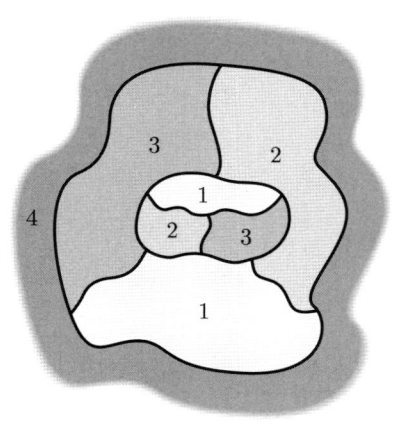

1.2 Only (d) is a map (although *not* a cubic map): (a) has a face with just two edges; (b) has a face that meets itself and a face (the exterior face) with just two edges; (c) has a vertex where only two edges meet and two faces with two edges in common.

1.3 The three colours that surround the vertex are not necessarily the same for each vertex: we may still need more than three colours altogether to colour the whole map. We cannot consider vertices in isolation.

1.4 **(a)** $V = 10$, $E = 15$, $F = 7$ (including the exterior face), and $V - E + F = 2$.
(b) $V = 20$, $E = 33$, $F = 15$, and $V - E + F = 2$.
(c) $V = 18$, $E = 27$, $F = 11$, and $V - E + F = 2$.

1.5 We have $F_3 = 2$, $F_4 = 4$, $F_5 = 3$, $F_8 = 1$, $F_9 = 1$, and (1.1) becomes $6 + 8 + 3 - 0 - 2 - 3 - 0 - \cdots = 12$.

1.6 **(a)** All terms other than F_3 and F_6 are zero, and (1.1) becomes $3F_3 = 12$, so $F_3 = 4$.
Thus, there are exactly four triangles.
(b) All terms other than F_5 and F_6 are zero, and (1.1) becomes $F_5 = 12$.
Thus, there are exactly twelve pentagons.

1.7 The dodecahedron is an example of a spherical map in which each face meets *exactly* five other faces.

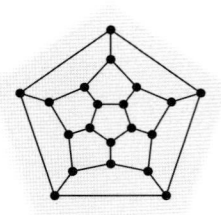

2.1 The graphs can be drawn in various ways — for example:

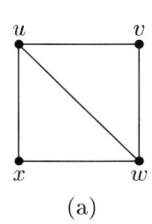

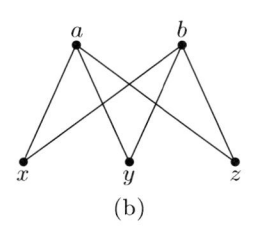

(a)

(b)

In graph (a), x is adjacent to u and w;
in graph (b), x is adjacent to a and b.

2.2 (a) One way of drawing them is as follows.

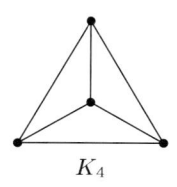

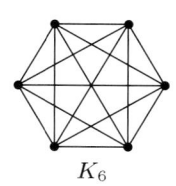

K_4 K_6

(b) K_n has $\frac{1}{2}n(n-1)$ edges.

2.3 (a) One way of drawing them is as follows.

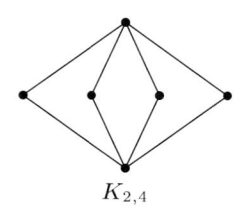

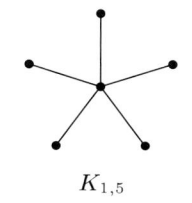

$K_{2,4}$ $K_{1,5}$

(b) $K_{r,s}$ has rs edges.

2.4 In K_n, each vertex has degree $n-1$.

In $K_{r,s}$, each of the vertices in the set of size r has degree s and each of the vertices in the set of size s has degree r.

2.5 Since each edge has two ends, it contributes exactly 2 to the sum of the degrees of the vertices.

Thus, the sum of the degrees of the vertices is equal to twice the number of edges.

(This result is sometimes called the *Handshaking Lemma* because, at a party, the sum of the numbers of guests with whom each guest shakes hands is twice the total number of handshakes.)

2.6 Only graphs (c) and (d).

The graph in the margin has no triangles or quadrilaterals; a pentagon is $vwxyzv$ and a hexagon is $vwxydav$.

2.7 Two possible plane drawings are the following.

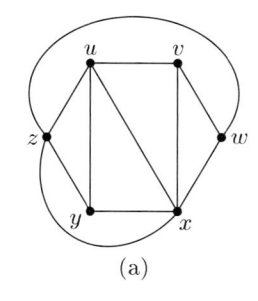

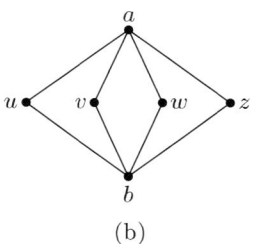

(a)

(b)

2.8 If there were a drawing of K_5 in the plane without crossings, then the cycle $uvwxyu$ in diagram (a) would have to appear as a pentagon. The edge vy would then lie either inside or outside the pentagon. Since the argument is similar in each case, we assume that vy lies inside the pentagon, as in diagram (b).

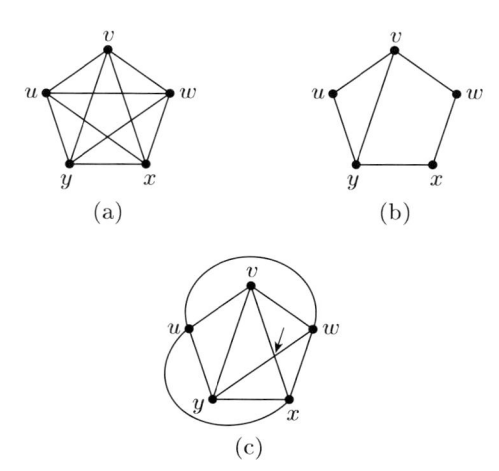

(a) (b)

(c)

Since the edges ux and uw cannot cross vy, they must both lie outside the pentagon, as in diagram (c). But the edge vx must not cross uw, and the edge wy must not cross ux, so both vx and wy must lie inside the pentagon, and must therefore cross (at the arrow). Since this is not allowed, it follows that K_5 has no plane drawing — that is, K_5 is non-planar.

2.9 (a) K_n is planar for $n \leq 4$; all other complete graphs contain K_5 and are thus non-planar.

(b) $K_{1,s}$ and $K_{2,s}$ are planar for all values of s; the diagrams below show $K_{1,5}$ and $K_{2,5}$.

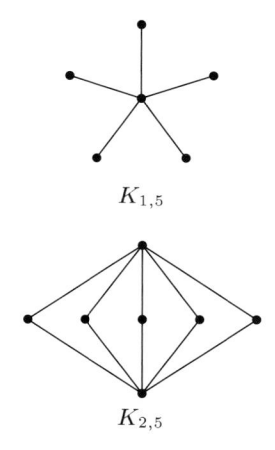

$K_{1,5}$

$K_{2,5}$

(c) At least one of r and s must be 1 or 2. All other complete bipartite graphs contain $K_{3,3}$, and are thus non-planar.

2.10 (a) 4; the diagram below shows a four-colouring.

(b) 2; the vertices can be coloured alternately with different colours.

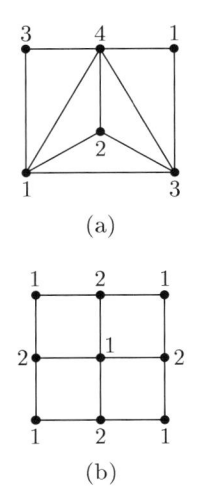

(a)

(b)

2.11 (a) n. (b) 2.
(c) 2 if n is even; 3 if n is odd.

2.12 Assume that the vertices of S^* can be coloured with k colours so that adjacent vertices are coloured differently. We now give each face of S the same colour as the vertex it contains: this produces a k-colouring of the faces of S. No two neighbouring faces of S can be coloured the same, since otherwise the vertices of S^* that they contain would be adjacent and have the same colour, which is not permitted. Thus, the faces of S can be coloured with k colours so that neighbouring faces are coloured differently.

3.1 One possibility is the following.

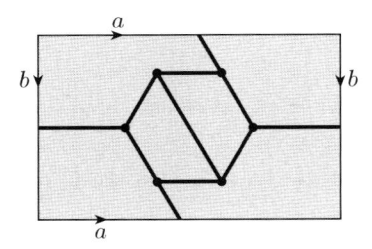

3.2 One possibility is the following.

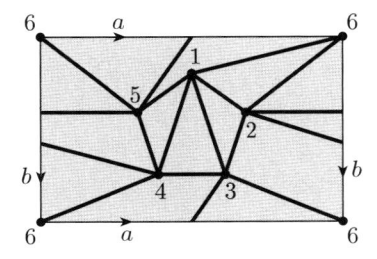

3.3 (a) One possibility is as follows.

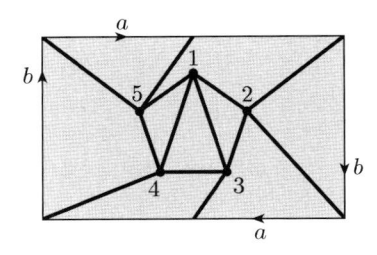

(b) One possibility is as follows.

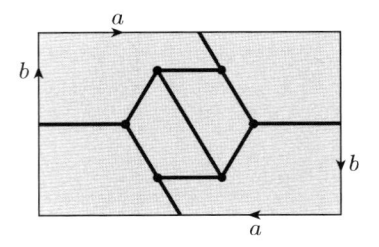

3.4 (a) $V = 7$, $E = 12$, $F = 5$,
and $V - E + F = 0$ ($= 2 - 2h$ when $h = 1$).

(b) $V = 6$, $E = 15$, $F = 9$,
and $V - E + F = 0$ ($= 2 - 2h$ when $h = 1$).

3.5 Since there are no triangles, each face has at least four edges. Each edge belongs to two faces, so, on counting edges, we obtain $2E \geq 4F$. Thus $F \leq \frac{1}{2}E$.

3.6 (a) $g(K_9) = \lceil \frac{1}{12}(9-3)(9-4) \rceil = \lceil \frac{30}{12} \rceil = 3$;
$g(K_{10}) = \lceil \frac{1}{12}(10-3)(10-4) \rceil = \lceil \frac{42}{12} \rceil = 4$;
$g(K_{11}) = \lceil \frac{1}{12}(11-3)(11-4) \rceil = \lceil \frac{56}{12} \rceil = 5$;
$g(K_{12}) = \lceil \frac{1}{12}(12-3)(12-4) \rceil = \lceil \frac{72}{12} \rceil = 6$;
$g(K_{13}) = \lceil \frac{1}{12}(13-3)(13-4) \rceil = \lceil \frac{90}{12} \rceil = 8$.

(b) Clearly the genus of K_n does not decrease as n increases. It follows from the above table that the smallest positive integer that is not the genus of any complete graph is 7.

3.7 $g(K_{5,5}) = \lceil \frac{1}{4}(5-2)(5-2) \rceil = \lceil \frac{9}{4} \rceil = 3$;
$g(K_{5,6}) = \lceil \frac{1}{4}(5-2)(6-2) \rceil = \lceil \frac{12}{4} \rceil = 3$;
$g(K_{5,7}) = \lceil \frac{1}{4}(5-2)(7-2) \rceil = \lceil \frac{15}{4} \rceil = 4$;
$g(K_{5,8}) = \lceil \frac{1}{4}(5-2)(8-2) \rceil = \lceil \frac{18}{4} \rceil = 5$.

3.8 K_5: $V = 5$, $E = 10$, $F = 6$,
and $V - E + F = 1$ ($= 2 - k$ when $k = 1$).
$K_{3,3}$: $V = 6$, $E = 9$, $F = 4$,
and $V - E + F = 1$ ($= 2 - k$ when $k = 1$).

3.9 (a) Suppose that G has non-orientable genus q, and that a simple embedding of G on the surface N_q has F faces, where $V - E + F = 2 - q$.

By Lemma 3.4(a), $F \leq \frac{2}{3}E$, and so
$$2 - q = V - E + F \leq V - E + \tfrac{2}{3}E = V - \tfrac{1}{3}E,$$
which can be rearranged to give $q \geq \lceil \frac{1}{3}(E - 3V + 6) \rceil$.

(b) If G has no triangles, then by Lemma 3.4(b) $F \leq \frac{1}{2}E$. Thus,
$$2 - q = V - E + F \leq V - E + \tfrac{1}{2}E = V - \tfrac{1}{2}E,$$
which can be rearranged to give $q \geq \lceil \frac{1}{2}(E - 2V + 4) \rceil$.

3.10 $q(K_7) = 3$;
$$q(K_8) = \lceil \tfrac{1}{6}(8-3)(8-4) \rceil = \lceil \tfrac{20}{6} \rceil = 4;$$
$$q(K_9) = \lceil \tfrac{1}{6}(9-3)(9-4) \rceil = \lceil \tfrac{30}{6} \rceil = 5;$$
$$q(K_{10}) = \lceil \tfrac{1}{6}(10-3)(10-4) \rceil = \lceil \tfrac{42}{6} \rceil = 7.$$

3.11 $q(K_{5,5}) = \lceil \tfrac{1}{2}(5-2)(5-2) \rceil = \lceil \tfrac{9}{2} \rceil = 5;$
$$q(K_{5,6}) = \lceil \tfrac{1}{2}(5-2)(6-2) \rceil = \lceil \tfrac{12}{2} \rceil = 6;$$
$$q(K_{5,7}) = \lceil \tfrac{1}{2}(5-2)(7-2) \rceil = \lceil \tfrac{15}{2} \rceil = 8;$$
$$q(K_{5,8}) = \lceil \tfrac{1}{2}(5-2)(8-2) \rceil = \lceil \tfrac{18}{2} \rceil = 9.$$

4.1 Suppose that such a map has V vertices, E edges and F faces.

Since a map is a simple embedding of a connected graph, we can combine Lemma 1.4 for maps with Theorem 3.3 for graphs. Thus, substituting the inequality $V \leq \frac{2}{3}E$ from Lemma 1.4(a) into Euler's Formula for a torus, we obtain
$$0 = V - E + F \leq \tfrac{2}{3}E - E + F,$$
which can be rearranged to give $E \leq 3F$.

On multiplying by $2/F$, we obtain:
$$\frac{2E}{F} \leq 6.$$

Since $2E/F$, the *average* number of edges per face, does not exceed 6, *at least one* face must have six or fewer edges. In other words, at least one face must meet six or fewer other faces.

4.2 The proof is by mathematical induction on the number of faces.

The result is clearly true for toroidal maps with up to seven faces.

We now assume that all toroidal maps with k faces can be coloured with at most seven colours, and show that all toroidal maps with $k + 1$ faces can be so coloured.

Let M be a toroidal map with $k + 1$ faces. By Theorem 4.2, M has a face A that meets at most six other faces. If we remove one of the edges e surrounding A (say the edge separating face A from face B), then the resulting map N has k faces, one of which (let us call it C) replaces faces A and B. By our induction hypothesis, we can colour the faces of the map N with seven colours in such a way that neighbouring faces are coloured differently.

We now reinstate the edge e, and colour M as follows. For the faces other than A and B, we use the colours of the corresponding faces of N. For face B, we use the colour that was used on face C of N. Now the only face left to colour is A. Since A has at most six neighbours, and seven colours are available, there is a spare colour that can be used. This gives a seven-colouring of the faces of the original map M.

The result now follows by mathematical induction.

4.3 $\text{ch}(O_3) = \lfloor \tfrac{1}{2}(7 + \sqrt{1 + 144}) \rfloor$
$$= \lfloor \tfrac{1}{2}(7 + 12.041...) \rfloor = \lfloor 9.520... \rfloor = 9;$$
$\text{ch}(O_4) = \lfloor \tfrac{1}{2}(7 + \sqrt{1 + 192}) \rfloor$
$$= \lfloor \tfrac{1}{2}(7 + 13.892...) \rfloor = \lfloor 10.446... \rfloor = 10;$$
$\text{ch}(O_5) = \lfloor \tfrac{1}{2}(7 + \sqrt{1 + 240}) \rfloor$
$$= \lfloor \tfrac{1}{2}(7 + 15.524...) \rfloor = \lfloor 11.262... \rfloor = 11;$$
$\text{ch}(O_6) = \lfloor \tfrac{1}{2}(7 + \sqrt{1 + 288}) \rfloor$
$$= \lfloor \tfrac{1}{2}(7 + 17) \rfloor = \lfloor 12 \rfloor = 12.$$

4.4 $\text{ch}(N_3) = \lfloor \tfrac{1}{2}(7 + \sqrt{1 + 72}) \rfloor$
$$= \lfloor \tfrac{1}{2}(7 + 8.544...) \rfloor = \lfloor 7.772... \rfloor = 7;$$
$\text{ch}(N_4) = \lfloor \tfrac{1}{2}(7 + \sqrt{1 + 96}) \rfloor$
$$= \lfloor \tfrac{1}{2}(7 + 9.848...) \rfloor = \lfloor 8.424... \rfloor = 8;$$
$\text{ch}(N_5) = \lfloor \tfrac{1}{2}(7 + \sqrt{1 + 120}) \rfloor$
$$= \lfloor \tfrac{1}{2}(7 + 11) \rfloor = \lfloor 9 \rfloor = 9;$$
$\text{ch}(N_6) = \lfloor \tfrac{1}{2}(7 + \sqrt{1 + 144}) \rfloor$
$$= \lfloor \tfrac{1}{2}(7 + 12.041...) \rfloor = \lfloor 9.520... \rfloor = 9.$$

4.5 When $k = 3$, $n = \lfloor \tfrac{1}{2}(7 + \sqrt{1 + 24k}) \rfloor = 7$. By Ringel's Theorem for $q(K_n)$ (Theorem 3.10), $q(K_7) = 3$. It follows that K_7 can be embedded on N_3, and hence that $\text{ch}(N_3) \geq 7$. Hence, when $k = 3$, we have
$$\text{ch}(N_k) \geq \lfloor \tfrac{1}{2}(7 + \sqrt{1 + 24}) \rfloor.$$
Combining this with Theorem 4.13 tells us that formula (4.2) holds for $k = 3$.

Index

adjacent edges, 18
Appel, Kenneth, 6

Birkhoff diamond, 51
Birkhoff, George, 51

chromatic number
    graph, 23
    surface, 39
colourable
    k-colourable, 23
colouring
    k-colouring, 23
Colouring Theorem
    for orientable surfaces, 41
connected graph, 20
cubic map, 10

degree of vertex, 19

edge of graph, 18
embedding, 26
Euler's Formula
    non-orientable surface N_k, 33
    orientable surface O_h, 29
    projective plane, 38
    spherical map, 11
    torus, 37

Face-counting Theorem, 12
Five-colour Theorem
    for planar graphs, 25
    for spherical maps, 15
four-colour problem, 6
Four-colour Theorem
    for planar graphs, 25
    for spherical maps, 8

genus
    of a non-orientable surface, 29
    non-orientable, 29
    of an orientable surface, 27
    orientable, 28
graph
    complete, 19
    complete bipartite, 19
    connected, 20
    edge, 18
    planar, 20
    simple, 18
    vertex, 18

Haken, Wolfgang, 6

Handshaking Lemma, 53
Heawood Conjecture, 6
Heawood's Formula, 39
hexagon, 13

k-colourable, 23
k-colouring, 23
Kempe, Alfred, 6

loop, 18

map, 10
    cubic, 10
multiple edges, 18

orientable
    genus, 28

pentagon, 11
planar graph, 17, 20
plane drawing, 20

quadrilateral, 11

reducible configuration, 51
Ringel's Theorem
    for $g(K_{r,s})$, 33
    for $q(K_n)$, 34
    for $q(K_{r,s})$, 35
Ringel, Gerhard, 6
Ringel–Youngs Theorem
    for $g(K_n)$, 32
    for orientable surfaces, 42

Seven-colour Theorem
    for toroidal maps, 37
simple embedding, 26
simple graph, 18
Six-colour Theorem
    for spherical maps, 14
spherical map, 8
subdivision
    finite, 10
subgraph, 19

triangle, 11

unavoidable set, 49

vertex of graph, 18
    degree, 19

Youngs, Ted, 6